THE PROALIDAE

Guides to the Identification of the Microinvertebrates of the Continental Waters of the World

Coordinating editor: Henri J. Dumont
State University of Gent, Belgium

ROTIFERA

Editor
Thomas Nogrady
Queen's University, Kingston, Ont. Canada

Editorial Committee
W. Koste, Quakenbrück, Germany
R. Shiel, Albury, Australia

SPB Academic Publishing bv 1996

Guides to the Identification of the Microinvertebrates of the Continental Waters of the World
Coordinating editor: H.J.F. Dumont

9

ISSN 0928-2440

ROTIFERA

Volume 4: The Proalidae (Monogononta)

by

Willem H. De Smet
University Centre of Antwerpen
Antwerpen, Belgium

SPB Academic Publishing bv 1996

CIP-DATA KONINKLIJKE BIBLIOTHEEK, DEN HAAG

Rotifera.

Rotifera / [ed. Thomas Nogrady]. – Amsterdam :
SPB Academic Publishing
Vol. 4: The Proalidae (Monogononta) / by Willem H. De Smet.
- Ill.- (Guides to the identification of the microinvertebrates
of the continental waters of the world, ISSN 0928-2440 ; 9)
With index, ref.
ISBN 90-5103-119-X
Subject headings: Rotifera.

ISBN 90-5103-119-X

Distributors:

For the U.S.A. and Canada:
SPB Academic Publishing bv
c/o Demos Vermande, Order Department
386 Park Avenue South, Suite 201
New York, NY 10016
Telefax (+212) 683-0118

For all other countries:
SPB Academic Publishing bv
P.O. Box 11188
1001 GD Amsterdam, The Netherlands
Telefax (+31.20) 638-0524

Table of contents

INTRODUCTION

The family Proalidae is a small group of monogonont Rotifera that was erected as the subfamily Proalinae of the Notommatidae by Harring & Myers (1924). It was given family rank by Bartoš (1959). The Proalidae comprises predominantly free-living epiphytic-epibenthic and psammobiontic species, inhabiting the littoral zone of fresh, thalassic and athalassic waters, and damp terrestrial habitats. Some species live epizoic on invertebrates and others are endoparasites of algae or ectoparasites of invertebrates.

Despite the early discovery of most of the species, the study of the Proalidae has been neglected due to the difficult identification of preserved specimens. As a consequence their distribution is insufficiently known and many species probably await description. The status of several species is confused, due largely to the fact that descriptions and illustrations of earlier authors are poor and only partially accurate. These descriptions mainly relied on the shape of the body, foot and toes, which are often rather similar in many members of the different genera. Less weight was given to the highly species-specific structure of the trophi which are, it must be admitted, very minute in some genera (10-20μm), and difficult to study with even modern techniques.

A thorough revision of the Proalidae is actually hampered by the above mentioned lack of adequate descriptions and all but inexistent type material. Moreover, apart from a few exceptions, most of the species are rare and collected infrequently, and comparatively few specimens were available for this study. It is, however, clear that only a comprehensive application of scanning electron microscopy for the study of trophi structure, will allow a better specific identification and understanding of the inter-relationship of the species.

ACKNOWLEDGEMENTS

I am greatly indebted to Mr A. Das, Mrs S. Pooters, Mr. J. Uytgeerts, Mrs S. Van Raemdonck and Dr E. Van Rompu for the help with typing and preparation of the illustrations. I am especially grateful to Ing. J. Van Daele for teaching me to use the S.E.M. and to the Department of Morphology (Cell Biology and Histology) for access to the S.E.M.. Dr G. Van Hassel is acknowledged for providing translations of Russian texts.

I am most grateful to Prof Dr H.J. Dumont who invited me to prepare this volume. I would like to thank Dr h.c. W. Koste, Prof Dr T. Nogrady and Dr R. J. Shiel for reading and commenting on the original draft of this Guide.

Finally sincere thanks also go to Dr. C. Friedrich for providing specimens of *Proales reinhardti* from arctic sea ice, Dr. C. Jersabek for providing specimens of *Proales doliaris* and Dr H. Segers for nomenclatorial remarks.

University of Antwerp
RUCA, Department of Biology
Groenenborgerlaan 171
B-2020 ANTWERPEN BELGIUM

FAMILY PROALIDAE HARRING & MYERS, 1924

Diagnosis

Illoricate, rarely semiloricate. Body swollen, fusiform or vermiform. Head, trunk and foot usually clearly defined. Corona simple, mostly supraoral, made up of the oblique buccal field and the lateral parts of the circumapical band. The trophi are malleate, virgate or a modification of one of these types. A hypopharyngeal muscle inserts on the floor of the mastax cavity and serves a pumping function. Eyespot(s) usually on brain, rarely frontal, lateral or absent.

Four genera and about 54 valid species: *Bryceella* (2), *Wulfertia* (3), *Proalinopsis* (6), *Proales* (43).

Morphology and internal organization

External morphology

The females of the Proalidae vary in length from 70μm in *Proales minima* to 1050μm in *P. segnis*. About 70% of the species are in the range of 100-250μm in length. The very few males that are known, attain smaller maximum lengths than the corresponding females.

The external form of the females varies much and may be vermiform, cylindrical, fusiform, vasiform or saccate, and dorso-ventrally flattened to various degrees, or with a flattened ventrum and arched dorsum. Depending on the species the outline of the body may be fairly constant or highly variable. The body is covered with a thin, very flexible, and usually smooth and very transparent cuticle, which may be slightly stiffened in some species.

Three main body regions may be distinguished: head, trunk and foot. In several species there is also a more or less differentiated neck. The different regions are separated from each other by more or less deep transverse folds in the cuticle. Each of the main regions can show pseudosegmentation by the presence of secondary transverse folds.

The head bears a simple, disc-shaped or oval corona with frontal, oblique or ventral orientation. The corona is made up of the buccal field and the lateral parts of the circumapical band in *Bryceella*, *Proalinopsis* and *Proales*, or reduced to a small circumapical band in *Wulfertia*. The mouth is at or near the ventral margin of the buccal field (*Bryceella*, *Proalinopsis*, *Proales*) or set some distance posterior to the reduced buccal field (*Wulfertia*). Several species of the genera *Proalinopsis* and *Proales* show two lateral tufts of densily set, long cilia at the sides of the buccal area, which resemble auricles but are not retractable. *Bryceella* is remarkable in having a corona with stout cirri, derived from cilia and placed in transverse rows, with which the animal moves in a jerky motion, and by the presence of two long sensory styli projecting from the head laterally. A very prominent projecting rostrum is developed from the apical field in *Bryceella*. Rostrum-like processes are also found in *Proales* (e.g. *P. macrura*, *P. decipiens*, *P. werneckii*). The dorsal antenna on the head or neck is usually inserted on a knob-like or cylindrical papilla in *Proalinopsis*; it is less prominent in the other genera.

The trunk constitutes the largest region and largely determines the general shape of the body. Many species show a distinct pseudosegmentation of the trunk (e.g. *Wulfertia ornata*, *Proales alba*, *P. provida*, *P. fallaciosa*, *P. decipiens)*, whereas others are characterized by one or more transverse dorsal or latero-dorsal folds (e.g. *Proales doliaris*, *P. cognita*, *P. gigantea*). In *Bryceella* and some *Proales* (e.g. *P. globulifera*, *P. minima*) the cuticle is weakly stiffened to thin dorsal and ventral plate-like sections separated by lateral clefts or sulci. The posterior end of the trunk is more or less abruptly narrowing (e.g. *Bryceella*, *Proales minima*, *P. halophila*) or gradually tapering into the foot (e.g. *Wulfertia*, *Proales adenodis*, *P. alba*, *P. ornata*). Many species display a tail taking the shape of a dorsal fold or short extension of the trunk, anterior to the anus (e.g. *Proales oculata*, *P. theodora*, *P. reinhardti*, *P. macrura*), which, however, is often hardly perceptible. The tail region of many *Proalinopsis* is provided with a dorsal papilla bearing a robust spine.

A foot is always present and extends from the body terminally. It may be short (*Wulfertia*) to very long (*Proales theodora*) and more or less parallel-sided (*Proales halophila*) or gradually tapering to the toes (*Proales alba*). It consists of one to four pseudosegments; the foot of *Proales similis* is usually wrinkled. *Proalinopsis caudatus* displays a setate dorsal papilla on the basal foot pseudosegment. In *Proales doliaris* the foot terminates in a single toe; the other Proalidae have two completely separated, or rarely partly fused, movable, symmetrical toes. The foot and toes are often telescopically retractable in the trunk. The toes are usually short to medium-sized, conical or lanceolate, straight or slightly decurved ventrally; in *Bryceella stylata* they show articulating joints. The toes of many species end in tubular points through which the pedal glands are secreting. Several species (e.g. *Proalinopsis caudatus*, *Proales palimmeka*, *P. pugio*, *P. cognita*, *P. gigantea*, *P. fallaciosa*, *P. ornata*) bear a dorsal papilla or spur of unknown function, between the insertion of the toes.

Internal organization

The internal organization of the Proalidae is of the same fundamental structure as in the monogonont rotifers throughout. Only those elements which are used in the keys and species descriptions are briefly discussed here. For more details on the subject, the reader is referred to Nogrady *et al.* (1993).

The head encloses the cerebral ganglion or brain and the retrocerebral organ. The brain is usually large and saccate. An unpaired retrocerebral sac may be present or lacking; it is often rudimentary. Paired subcerebral glands have been reported in a few marine species only (*Proales oculata*, *P. halophila*, *P. germanica*, *P. paguri*, *P. christinae*), but went probably unnoticed in many species.

Most species have one or two red eyespots, with or without crystalline bodies. The red pigment is occasionally lacking or disappearing in adults (e.g. *Proales gigantea*, *P. daphnicola*). Most species (*Wulfertia*, *Proalinopsis*, *Proales*) have a single eyespot, or less often, two cerebral eyes (a large right and a smaller left) at the dorsal or ventral edge of the brain. The cerebral eye(s) are often displaced to the right. Some species have two close-set or fused frontal eyespots located in the corona (e.g.

Proales globulifera, P. minima, P. commutata, P. oculata, P. gonothyraeae, P. theodora, P. reinhardti). The lateral light-refracting bodies at the base of the styli in *Bryceella stylata* have been regarded as eyes by some authors.

The digestive tract begins with the mouth-opening which leads to the pharynx, part of which is transformed into the muscular mastax containing the jaws or trophi. It is continued into the oesophagus, stomach and intestine, and terminates in the anus which opens dorsally, posterior to the foot. Stomach and intestine may (e.g. *Proales doliaris*) or may not (e.g. *P. oculata*) be separated by a more or less deep transverse constriction. The stomach bears near the entrance, a pair of gastric glands of variable size and shape. The gastric glands are usually short-stalked (with long stalks in *Proales syltensis*). In *P. syltensis*, *P. halophila* and *P. reinhardti* the stomach is preceeded by an enlarged section of the oesophagus, the proventriculus. A glandular caecum, attached posteriorly to the ventral wall of the intestine, has been reported in *Proales segnis*.

The trophi are modified malleate, virgate or intermediate between malleate and virgate. The hypopharyngeal muscle is inserted on the wall of the mastax cavity and not on the fulcrum; it produces a pumping action. The generally triangular rami may be massive or thin, flat or curved, symmetrical or strongly asymmetric. Their inner margins may be toothed; large basal apophyses are not uncommon and alulae are often well-developed. The fulcrum is more or less short and nearly in a straight line with the rami, or attached at a more or less great angle to the axis of the rami. It generally points towards the posterior rather than ventral. The unci have 1-8 large teeth; the principal tooth may be bifurcate, or provided with a small accessory toothlet or knob-like prominence on the ventral margin; at the dorsal margin a group of 3-4 small toothlets may be distinct. The manubria consist of the three longitudinal chambers or cavities, which form the head anteriorly and the handle posteriorly; the lateral chambers may be transformed into plates or lamellae. It is not always possible to distinguish lateral chambers or cavities from lamellae by light microscopy; these lateral structures are therefore referred as lamellae in the descriptions of the trophi, in accordance with the earlier literature. The manubria are variable: long, slender, strongly incurved posteriorly and head with 2 small lateral lamellae, to short, broad, straight and with broad and long lateral lamellae. The epipharynx is present and variable: two serrate plates, a dome-shaped lamella or two club-shaped, sigmoid, rod-shaped,... pieces, etc.

A more detailed analysis of trophi-structure in each genus based on S.E.M. pictures is to be found with the descriptions of the species.

Other internal characters that are sometimes of importance in identification are size and shape of the pedal glands, and the presence of accumulations of light-refracting granules in the body cavity (e.g. *Proales globulifera*). The pedal glands may be very long and extending into the trunk, with reservoirs in the foot pseudosegments and toes.

The internal organization of the males of the Proalidae is insufficiently known. Males of *Proales reinhardti* and *P. werneckii* apparently have a functional mastax and rudimentary digestive tract.

Eggs are only described for a few species. Subitaneous eggs are usually oval to rounded and smooth. In *Proales syltensis* and *P. christinae* they are provided with a short stalk, by which they stick to the substrate. The eggs of *P. halophila* (*sensu* Tzschaschel, 1979) are more or less rectangular with concave underside and convex upper side; they are attached at their concave side to sand-grains, detritus etc., with the four, slightly drawn out corners. Resting eggs are thick-shelled, smooth (*Proales werneckii*), coarsely granulated (*P. daphnicola*) or equipped with spines (*P. parasita*).

Distribution and ecology

The Proalidae as a group show a cosmopolitan distribution. Zoogeographic observations on the different species are very limited, and many of them have not been reported since their description. They were most intensively studied only for N. America and W. Europe, however, data on species composition and ecology are still very poor for this regions indeed.

The Proalidae are found in diverse habitats within freshwater, inland saline, brackish and marine environments. *Bryceella*, *Wulfertia* and *Proalinopsis* are exclusively found in freshwater habitats. Eleven of the 43 *Proales* species inhabit saline waters. *Proales theodora* is unusual in that it occurs in mountain springs, the littoral of alkaline stagnant and running fresh waters, and among algae in the littoral of brackish and marine waters. *Proales similis* displays an extreme salinity tolerance: it is encountered in brackish and marine waters, and in hypersaline springs up to a salinity of 98‰, or almost three times the concentration of sea water. Most of the species live in the littoral zone of ponds, lakes and seas. They are predominantly periphytic and epibenthic; few freshwater and many marine species are also psammobiontic. Some marine species (*Proales syltensis*, *P. commutata*, *P. germanica*, *P. oculata*) appear to be restricted to the sand habitat. *Bryceella stylata* is typical of wet terrestrial habitats and commonly found in wet moss cushions, among leaf litter, in tree holes, etc.; it is also found among mosses in acid waters.

The Proalidae, especially the genus *Proales*, are fairly outstanding among the rotifers in exhibiting a variety of relations to other organisms, varying from epizoic associations with animals to true parasites of algae and invertebrates. *Proales daphnicola* lives on the carapace of *Daphnia* spp. and other crustaceans (copepods, amphipods, etc.), where it feeds on epizoic unicellular euglenaceans (*Colacium*) and ciliates; it also attaches its eggs to the carapace. *Proales sigmoidea* lives among sessile colonial ciliates (e.g. Epistylidae, Vorticellidae) whereupon it also feeds. *Proales paguri* was found on the gills of the hermit crab *Eupagurus bernhardus* (L.), where it was seen browsing on the gills and succing the hosts epithelium. *Proales gonothyraeae* and *P. christinae* are associated with hydroid polyps; the former lives and lays eggs inside the theca of *Laomedea loveni* (Allm.) and is considered to be an ectoparasite that feeds on the epidermal cells of its host. *Proales decipiens* usually feeds on detritus, bacteria and small algae, but has also been found feeding on adults and developing eggs of the rotifer *Stephanoceros fimbriatus* (Goldf.). The eggs of several freshwater gastropods are parasitised by *Proales gigantea*; the young female enters the snail egg by piercing a hole through the egg shell, and feeds on the fluid surrounding the embryo. Adult rotifers lay 7-13 eggs, one at a time, inside the egg; the first juvenile to hatch feeds on the remains of the snail egg, before escaping and infecting new snail eggs; those which hatch later tend to starve. When an egg-laying adult *P. gigantea* dies inside the snail egg, it is eaten by her progeny. *P. gigantea* has also been reported ovipositing in egg masses of chironomids and feeding on their eggs.

Colonial freshwater algae such as *Volvox* spp., *Uroglena* sp. and *Uroglenopsis-americana* (Calk.) are parasitised by *Proales parasita*. The rotifers penetrate into the interior of the colonies where they feed on the colonial cells and breed. *Proales parasita* has also been reported to be parasitic on the colonial ciliate *Ophrydium*. The filamentous algae *Vaucheria* spp. (both aquatic and terrestrial species) and *Dichotomosiphon tuberosus* (Braun) are parasitised by *Proales werneckii*. The rotifer enters the siphonous filaments and induces the formation of a gall in the form of a variously shaped, often lobed outgrowth. The rotifers feed on the contents and dy after depositing up to 80 eggs in the gall. The developing eggs hatch and the young escape to parasitise new filaments; copulation is outside the host; resting eggs overwinter in the sediment.

Very little is known about the feeding habits of free-living Proalidae. *Bryceella stylata* and *Proales globulifera* eat o.a. small diatoms; diatoms also, as well as dinoflagellates and bacteria are on the menu of *Proales reinhardti* and *P. theodora*; *P. fallaciosa* is feeding on detritus, bacteria and algae, and acts as scavenger, cleaning out the contents of dead cladocerans, copepods, insect-larvae, oligochaetes etc.

Key to genera

1 Corona with long cirri; a long stylus directed to the posterior, on each side of the head; head with rostrum (figs. 1-16) .... **1. *Bryceella*** (p. 7)
– Corona without long cirri; head without styli; rostrum if present weakly developed 2

2 Corona reduced, without lateral ciliary tufts; mouth set some distance posterior to the buccal field; toes small and stubby (figs. 17-39) .... **2. *Wulfertia*** (p. 11)
– Corona not reduced, usually with lateral tufts of long cilia; mouth in buccal field 3

3 Tail region or basal foot pseudosegment with papilla bearing a tuft of setae or a spine; dorsal antenna a small setigerous papilla, or a conspicuous projection; toes slender and acute (figs. 40-72) .... **3. *Proalinopsis*** (p. 18)
– Without such a setae- or spine-bearing papilla; dorsal antenna never a distinct projection; toes (one species with a single toe) usually ± short (figs. 73-328) .... **4. *Proales*** (p. 26)

1. Genus *Bryceella* Remane, 1929

Bryceella Remane (1929-33), Bronn's Klassen und Ordnungen des Tier-Reichs, Leipzig 4: 114-115. - Type species: *Bryceella stylata* (Milne, 1886), figs. 1-5.

Diagnosis

Body fusiform, slender; cuticle relatively stiff. Head region depressed, offset from trunk by distinct constriction; rostrum present; a very long, lateral stylus at each side of the head. Trunk oval, laterally with longitudinal sulci. Tail short, blunt. Foot relatively slender, with 2-3 pseudosegments. Toes slender, slightly decurved ventrally, bluntly pointed. Corona ventral, an oval disc with long cirri, placed in transverse rows; cirri increasing in length anteriorly; the cirri are used for walking.

Trophi modified malleate, small, symmetrical. Rami simple, trapezoid, with or without teeth on their inner margins as well as near the basal apophyses. Fulcrum short, rod-shaped in ventral view. Unci with 5-7 teeth. Manubria with lateral lamellae.

Two species.

Wide-spread.

Key to species

– With light-refracting bodies at base of styli; toes forceps-shaped, curved inwardly in dorsal view, with 3 articulating joints; length/width ratio terminal foot pseudosegment c. 2/1; rami with teeth on inner margins (figs. 1-5) .... **1. *B. stylata***

– Without light-refracting bodies at base of styli; toes curving outwardly in dorsal view, without articulating joints; length/width ratio terminal foot pseudosegment c. 3/1; rami without teeth on inner margins (figs. 6-11) .... **2. *B. tenella***

Descriptions

1. *Bryceella stylata* (Milne, 1886)

Figs. 1-5, pl.1 figs. 1-5

Milne 1886: 143, pl. 2 figs. 3,4,9 (*Stephanops stylatus)*; Harring 1913: 97 (*Squatinella stylata*); Remane 1929-33: 544, fig. 100 (*Bryceella stylata*); Wulfert 1940: 561-562, figs. 1a-f; Donner 1952b: 32, fig. 3; Koste 1968a: 125, figs. 2a,b; Kutikova 1970: 483, figs. 690a-f; Koste 1978: 264, pl. 87 fig. 10, pl. 88 figs. 2a-f; Koste & Shiel 1990: 130-131, figs. 1a-c; Jersabek & Schabetsberger 1992a: 100, figs. 41a-c.

Type locality

In moss near Glasgow, Scotland.

Description

Body fusiform, dorso-ventrally flattened; cuticle somewhat stiffened; colourless, hyaline. Head distinctly offset by neck-folds, subsquare; rostrum broad. Two long, straight, lateral styli, with light-refracting bodies at their base. Trunk oval, without pseudosegments; tail broad, semi-circular. Foot relatively short, broad, with 2(3) pseudosegments, length/width terminal foot pseudosegment c. 2/1. Toes forceps-shaped, slightly curved inwardly, with 3 articulating joints.

Trophi small, modified malleate. Rami with trapezoid dorsal surface and triangular median surface (with respect to the median plane of the incus). Dorsal surfaces with 2 short, blunt projections anteriorly and short points posteriorly; inner margins

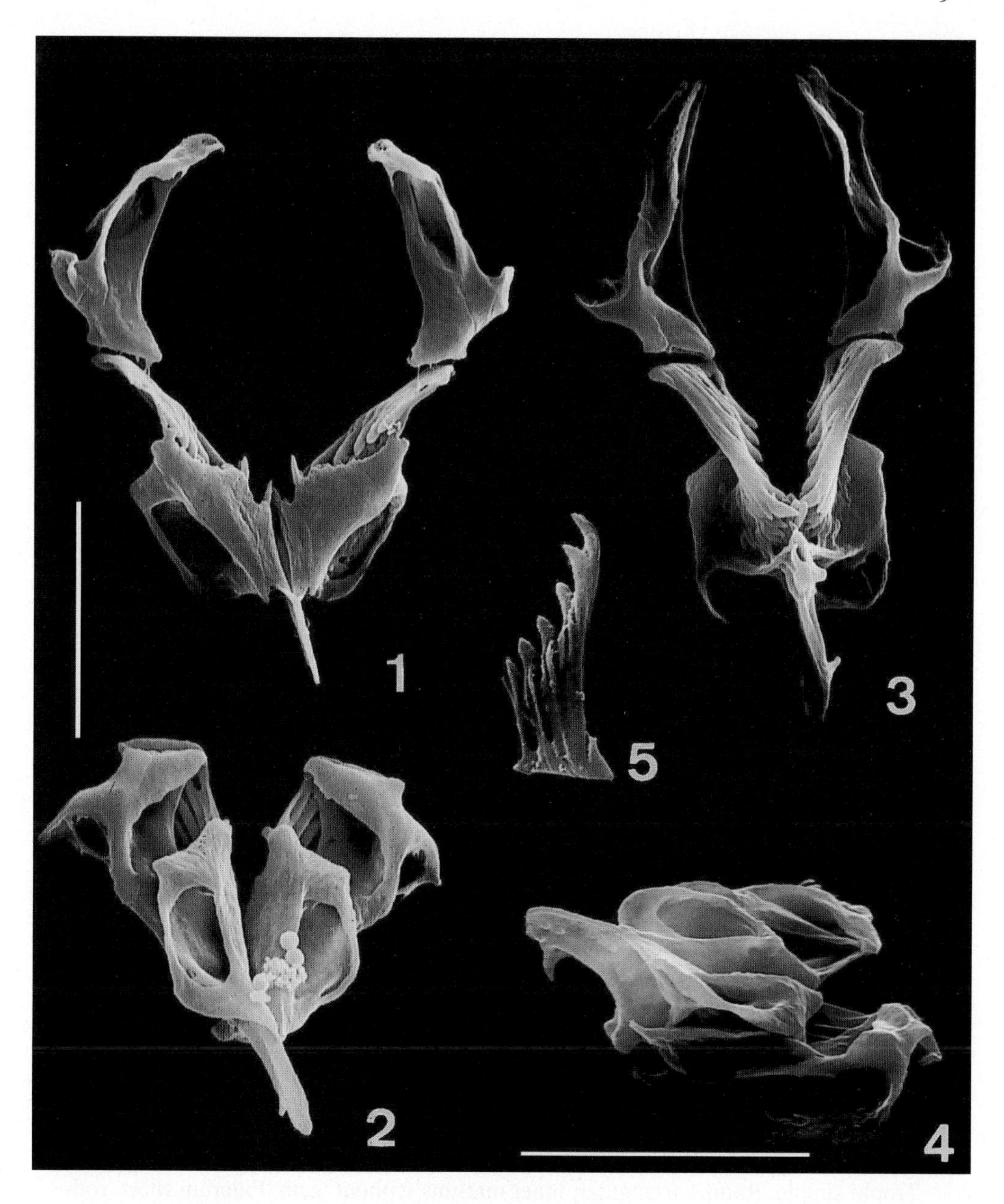

Plate 1. *Bryceella stylata*, trophi (S.E.M. photographs). 1,2: dorsal view, 3: ventral view, 4: lateral view, 5: right uncus.
Scale bars: 10µm.
(1-5: Canigou, Pyrenées Orientales, France)

of median surfaces with two stout bifurcate spines posteriorly, anteriorly ± serrate; large dorsal and basal fenestrae; two close-set apophyses near base of fulcrum; a small, striated plate with serrate margin between unci and rami, often projecting beyond rami tips; right a small chain of 6 ± lenticular elements. Fulcrum short, rod-shaped in dorsal view, in lateral view ± wedge-shaped, with ventral hook-shaped projection posteriorly. Unci 7-toothed, elongate. Manubria broad, slightly curved, each with long posterior and short anterior cavity.

Length 110-186 μm, width 40-80 μm, foot 18-28 μm, toe 12-21 μm; trophi 17-26 μm: ramus 8-9 μm, fulcrum 4-5 μm, uncus 8-10 μm, manubrium 9-12 μm.

Distribution and ecology
Palaearctic, N. America (Florida ; Canada, Devon Island; Alaska). In moor and acid waters between *Sphagnum*, also common in leaf litter and tree holes; more frequently in the cold season; 2-20 °C, pH 4.5-8.44, 167-232 μScm^{-1}, 2.9-5.6 mg $O_2 l^{-1}$. Food: small diatoms.

2. *Bryceella tenella* (Bryce, 1897)

Figs. 6-11

Synonym: ? *B. agilis* Neiswestnowa-Shadina, 1935

Bryce 1897: 798-799 (*Stephanops tenellus)*; Harring 1913: 97 (*Squatinella tenella*); Remane 1929-33: 114, 545, fig. 100 (*Bryceella tenella*); Wiszniewski 1932: 90-91 (male); Wiszniewski 1934b: 343-344, pl. 58 figs. 2-5; Wiszniewski 1935: 223-225; ? Neiswestnowa-Shadina 1935: 559-560, fig. 4 (*B. agilis*); Koste 1968a: 125, fig. 4; Kutikova 1970: 483, figs. 689a-d; Koste 1978: 264-265, pl. 87 figs. 9a-c, pl. 88 figs. 1a-e.

Type locality
In mosses from Spitsbergen, Svalbard.

Description
Body fusiform, flattened dorso-ventrally; cuticle delicate; colourless. Head prominent, subsquare, distinctly offset by neck-fold; rostrum broad. Two long, straight, lateral styli, without light-refracting bodies at their base. Trunk oval, without pseudosegments; tail broad, semi-circular. Foot slender, retractable, 2 pseudosegments, length/width terminal foot pseudosegment c. 3/1. Toes rather stiff, decurved outwardly in dorsal view, bluntly pointed, retractable.

Trophi simple. Rami ± triangular, inner margins without teeth. Fulcrum short, rod-shaped in ventral view. Unci 5-toothed. Manubria slightly curved, with long lamella.

Length 90-180 μm, toe 11-15 μm; trophi 15-17 μm.

Male (fig. 8) similar in habitus as female. Length 140 μm, toe 12 μm.

Distribution and ecology
Cosmopolitan. Among *Sphagnum* and in psammon of acid waters.

Species inquirendae

Bryceella agilis: Neiswestnowa-Shadina, 1935: 559-560, fig. 4. (figs. 12,13).
Body somewhat short, pyriform. Head margin with lobate projections anteriorly; styli short.

Length 105 µm, foot 33 µm, toe 12 µm.

Russia, in psammon of River Oka.

Synonymous with *B. tenella* according to Wiszniewski (1935).

Bryceella voigti: Rodewald, 1935: 84, figs. 3a,b. (figs. 14-16).
Body fusiform; cuticle rather stiff; bright blue-greenish, locally colourless. Head offset by neck-fold; lateral styli apparently absent; light-refracting bodies present. Trunk with pseudosegments; tail short, blunt. Foot long, slender, 2 pseudosegments, length/width terminal foot pseudosegment c. 4/1. Toes slender, acute, not retractable.

Trophi small, description insufficient.

Length 80-130 µm, width 37-42 µm, foot 14-26 µm, toe 6-12 µm.

Europe (Romania, British Isles), Australia. In *Sphagnum* pools and moss on *Eucalyptus* trunks; summer; pH 4.4.

Figures and description unsatisfactory. The status of this taxon was queried by Koste (1978) and Koste & Shiel (1990).

2. Genus *Wulfertia* Donner, 1943

Wulfertia Donner (1943), Zool. Anz. 143: 30-32, figs. 8a-h.- Type species: *Wulfertia ornata* Donner, 1943, figs. 17-30.

Diagnosis

Body ± fusiform, illoricate. Head indistinctly offset from trunk. Corona almost frontal, reduced, a simple, small circumapical band, uniformly covered with short cilia; mouth set some distance posteriorly, outside buccal field. Tail small. Foot very small. Toes short, acute. Cerebral eyespot(s) placed to the right. Brain saccate. Retrocerebral organs absent. Vitellarium with 8 nuclei. Bladder present.

Trophi virgate. Unci with (3)4-5 large teeth. Manubria broad. Epipharynx and pleural rods present.

Three species.

Eurasia, N. America, Africa, Arctica, Antarctica.

Key to species

1 Cuticle with longitudinal folds; rami ± triangular, without pronounced alulae (figs. 17-30) .... **1. *W. ornata***
– Cuticle without longitudinal folds; rami ± sickle-shaped, each with a pair of pronounced alulae 2

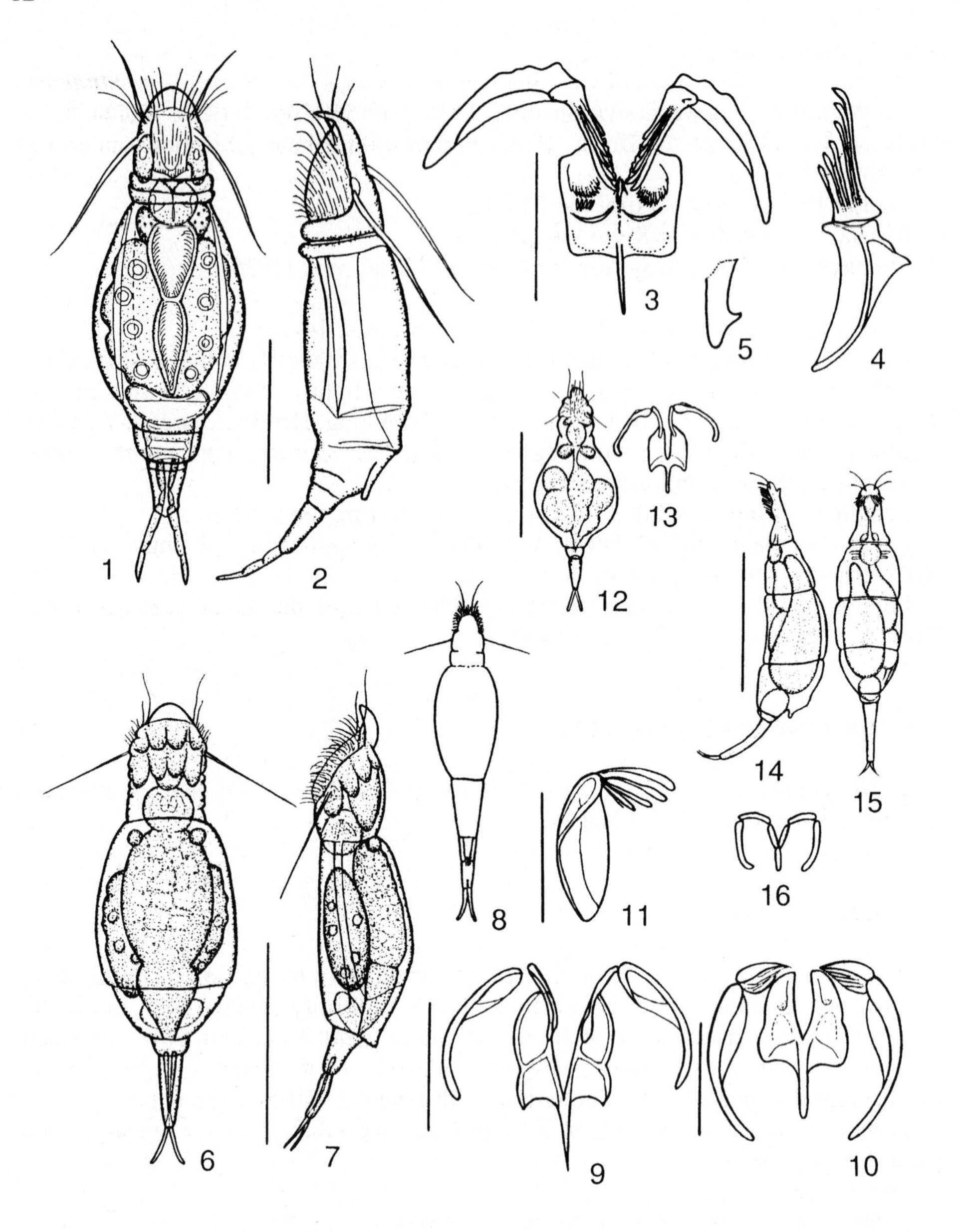

Figs. 1-5: *Bryceella stylata*. 1: dorsal view, 2: lateral view, 3: trophi, ventral view, 4: malleus, lateral view, 5: fulcrum, lateral view.
Figs. 6-11: *Bryceella tenella*. 6: dorsal view, 7: lateral view, 8: male, dorsal view, 9,10: trophi, ventral view, 11: malleus, lateral view.
Figs. 12,13: *Bryceella agilis*. 12: dorsal view, 13: trophi.
Figs. 14-16: *Bryceella voigti*. 14: lateral view, 15: dorsal view, 16: trophi.
Scale bars: habitus 50 μm, trophi 10 μm.
(1,2: after Wulfert, 1940; 3-5: Canigou, Pyrenées Orientales, France; 6,7,9,11: after Wiszniewski, 1935; 8: after Wiszniewski, 1934a; 10: after Koste, 1968a; 12,13: after Neiswestnowa-Shadina, 1935; 14-16: after Rodewald, 1935)

2 Pseudorostrum present; trunk without pseudosegments; anterior and posterior alulae narrow, tapering and curved (figs. 31-33) .... **2. *W. kindensis***
– Pseudorostrum absent; trunk with pseudosegments; anterior pair of alulae broad, rounded or with short point laterally; posterior pair narrow, tapering and curved (figs. 34-37) .... **3. *W. kivuensis***

Descriptions

1. *Wulfertia ornata* Donner, 1943

Figs. 17-30, pl. 2 figs. 1-8

Donner 1943: 30-32, figs. 8a-h; Wulfert 1960: 329-330, figs. 58a-f; Koste 1978: 265, pl. 88 figs. 4a-f, 5a-h; Kutikova 1970: 483-484, figs. 691a-d; De Smet 1993: 15, figs. 26a-d.

Type locality
Stone-pits near Kirschfeld, Znaim, Czechia.

Description
Body elongate fusiform, the head and foot merge into it; cuticle delicate; very hyaline. Head region sometimes with transverse dorsal folds. Rostrum absent. Posterior part of trunk ± widened at 1/3 from its end; median part often with pseudosegments; trunk dorsally with longitudinal furrows. Tail short. Foot small, short, one pseudosegment, slightly incised between toes. Toes small, stubby. Mouth surrounded by symmetrically oriented folds. Brain large, saccate. A large, red eyespot on brain, placed to the right; a smaller pigmentspot left of main one. Stomach and intestine separated by constriction. Gastric glands rounded with conspicuous nuclei. Pedal glands pyriform.

Trophi virgate, symmetrical. Rami ± trapezoid in dorsal view with plate oriented towards median plane of incus, slightly curved dorsally; rami tips with 3-4 small, blunt teeth; apically near inner margins with 3-5 broad teeth; dorsal fenestra ± small near posterior margin; a small, rounded ventral fenestra and elongate fenestra near attachment of fulcrum. Fulcrum medium long, ± rod-shaped in ventral view; in lateral view with broad base, slightly tapering, slightly expanded ventrally in posterior half. Unci with 4-5 strong teeth, gradually decreasing in size, and group of 3-4 smaller toothlets, principal teeth with accessory toothlet near tip; inner side anteriorly with skleropili and brush-like formations. Manubria broad, median cavity very large, 2 short, small lateral cavities; posterior end curved dorsally. Epipharynx two plates with serrate anterior margins; pleural rods elongate, wedge-shaped.

Length 102-180 μm, toe 6-9 μm; trophi 16-30 μm: ramus 9-11 μm, fulcrum 7-9 μm,, uncus 10-12 μm, manubrium 17-23 μm, epipharynx plate 5-10 μm.

Distribution and ecology
Europe (Czechia, Central Germany, Romania, Belgium, Svalbard), Africa (Kenya), N. America (Canada, Victoria Island). Among aquatic vegetation in ponds and shallow waters; April-November; pH 6.0-8.45, 367 μScm^{-1}.

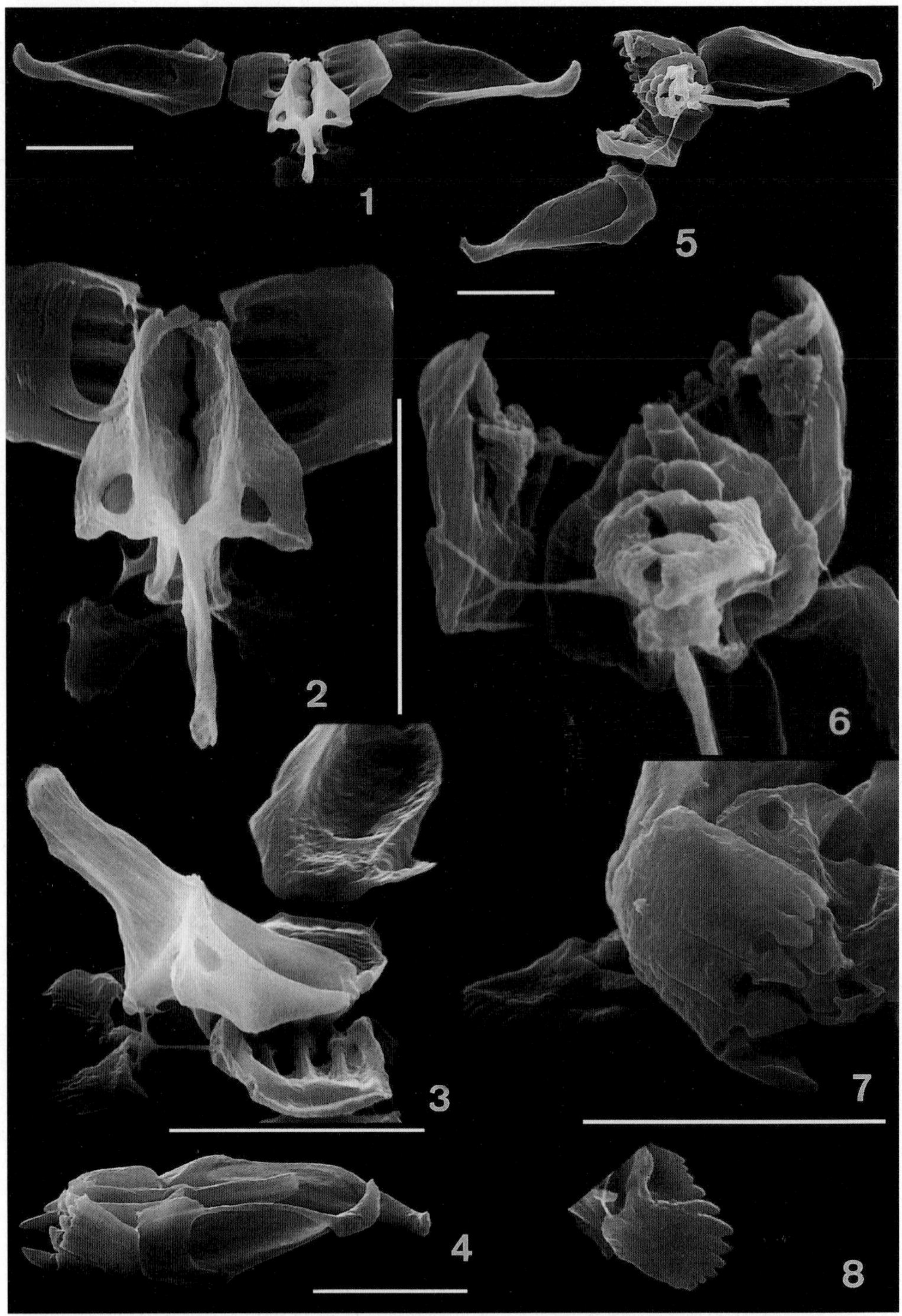

Plate 2. *Wulfertia ornata*, trophi (S.E.M. photographs). 1: dorso-caudal view, 2: idem, detail incus, 3: lateral view, detail incus, 4: lateral view, 5: ventro-apical view, 6: idem, detail incus, 7: detail right uncus with epipharyngeal plate, 8: epipharyngeal plate. Scale bars: 10μm. (1-4: Antwerpen, Belgium; 5,6: Victoria Island, N.W.T., Canada; 7-8: Barentsøya, Svalbard)

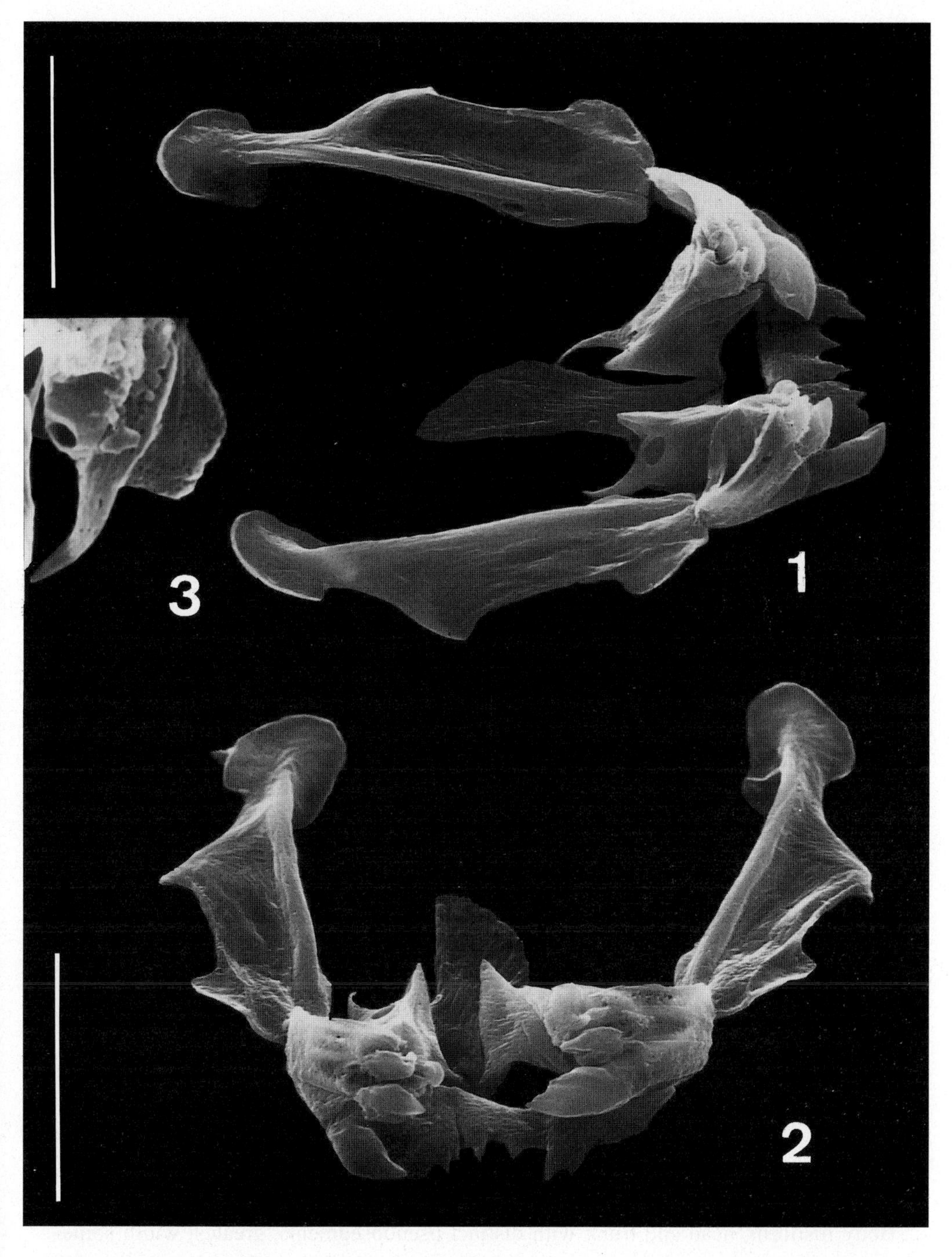

Plate 3. *Wulfertia kivuensis*, trophi (S.E.M. photographs). 1: ventral view (manubria inner side), 2: apical view, 3: detail right alulae.
Scale bars: 1,2: 10μm, 3: 7μm.
(1-3: Lake Kivu, Zaire)

2. *Wulfertia kindensis* Koste & Tobias, 1990

Figs. 31-33

Koste & Tobias 1990: 103-106, figs. 17a-c.

Type locality and type

Kinda-reservoir on Panlaung Chaung, left hand tributary to river Irrawaddy in central Myanmar (Burma). Holotype in the Senckenberg Museum and Forschungsinstitut, Frankfurt a. Main, Germany, SMF GP Rot 7305.

Description

Body squat, fusiform; cuticle soft. Head narrowed anteriorly, with pseudorostrum. Foot blunt conical, one pseudosegment. Toes very short, conical, acute, tips slightly decurved. Brain saccate. Retrocerebral glands absent. Eyespot red, on brain, placed to the right. Salivary glands apparently absent. Gastric glands large, rounded, nucleated. Stomach cellular, separated from intestine by constriction. Pedal glands large, with reservoirs.

Trophi resembles malleate, used virgate, relatively large. Rami ± sickle-shaped, tips bluntly toothed; each ramus with a pair of large, narrow, curved and tapering alulae. Fulcrum rod-shaped in ventral view, laterally broad rectangular. Unci 5-toothed. Manubria broad, expanded anteriorly, crutched posteriorly, with long lateral lamellae. Two epipharyngeal plates with serrate anterior margin. Two pleural rods.

Length 90-130 μm, toe 4-5 μm; trophi 35 μm: ramus 11.5 μm, fulcrum 14 μm, uncus 11 μm, manubrium 24 μm.

Distribution and ecology

Central Burma. In periphyton of man-made lake; pH 7.8-8.5, 230-430 μScm^{-1}.

3. *Wulfertia kivuensis* De Smet, 1992: n. stat.

Figs. 34-37, pl. 3 figs. 1-3

Synonym: *W. kindensis kivuensis* De Smet, 1992

De Smet *in* De Smet & Bafort 1992: 112-114, figs. 1a-e.

Type locality and types

Lake Kivu, Katana district, east from mission hospital Fomulac, 1.462 m a.s.l., Zaïre. Holotype in the Koninklijk Belgisch Instituut voor Natuurwetenschappen, Brussels, Belgium, No. AI 28.013; paratypes in R.U.C.A.

Description

Body fusiform, head and trunk with distinct pseudosegments, greatest width somewhat before mid-length. Head narrowed anteriorly, without pseudorostrum. Pseudosegments of trunk with delicate transverse striae ventrally, dorsally longitudinal ones. Foot blunt, conical, one pseudosegment. Toes very short, outer edges slightly curved, inner edges swollen at bases and slightly outcurved near mid-length, tips

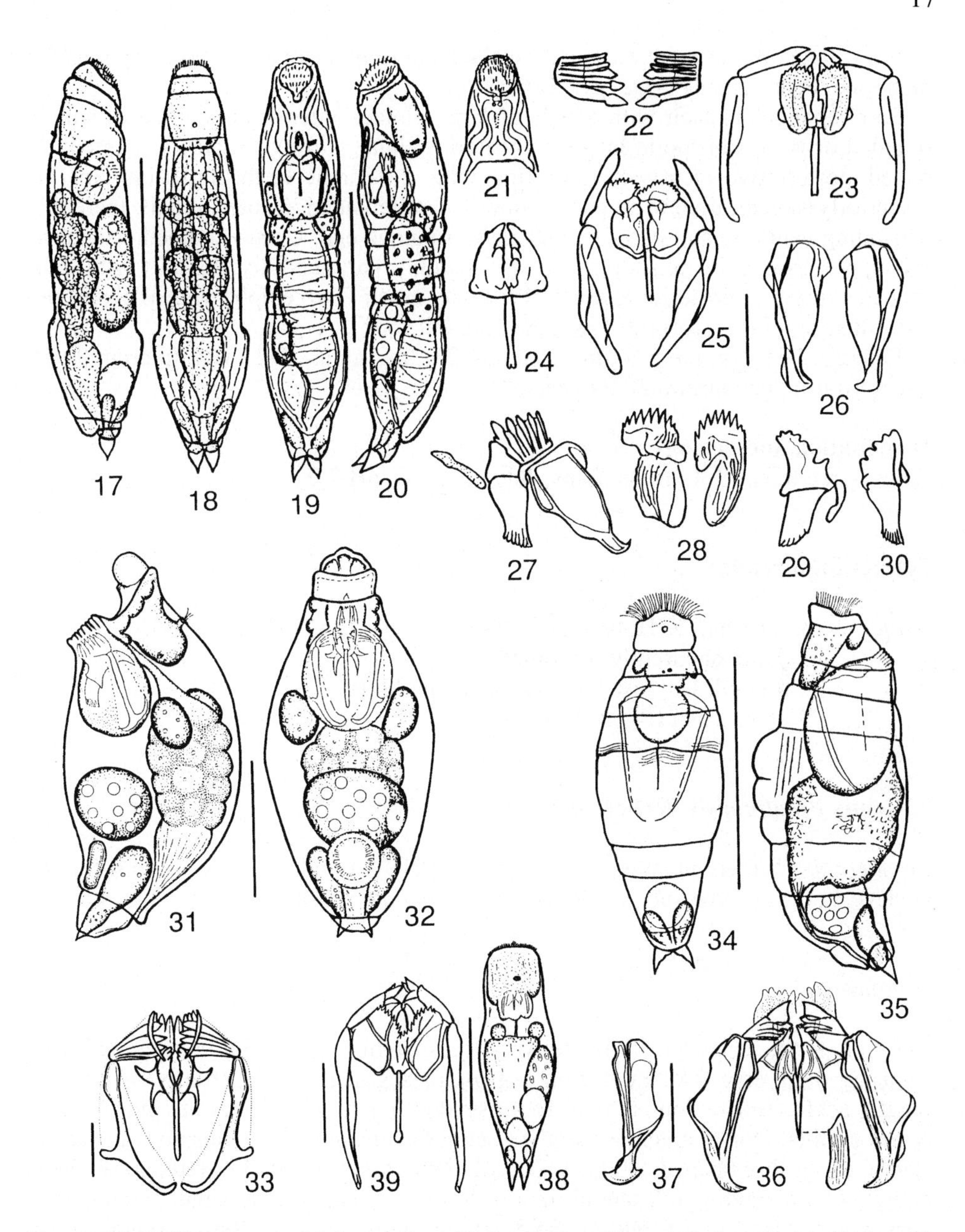

Figs. 17-30: *Wulfertia ornata*. 17: lateral view, 18: dorsal view, 19: ventral view, 20: lateral view, 21: head, ventral view, 22: unci, 23: trophi, ventral view, 24: incus, ventral view, 25: trophi, ventral view, 26: manubria, 27: trophi, lateral view, 28: epipharynx, 29,30: incus, lateral view.
Figs. 31-33: *Wulfertia kindensis*. 31: lateral view, 32: ventral view, 33: trophi, ventral view.
Figs. 34-37: *Wulfertia kivuensis*. 34: dorsal view, 35: lateral view, 36: trophi, ventral view ; fulcrum, lateral view, 37: manubrium.
Figs. 38,39: *Wulfertia* sp. 38: ventral view, 39: trophi.
Scale bars: habitus 50 μm, trophi 10 μm.
(17,18,21,23,24,27,28,30: after Donner, 1943; 19,20: after Wulfert, 1960; 22,25,26,29: after De Smet, 1993; 31-33: after Koste & Tobias, 1990; 34-37: after De Smet, 1993; 38-39: after Dartnall & Hollowday, 1985)

acute. A large, red eyespot on brain, displaced to right, a smaller pigmentspot left of main one. Pedal glands large, pyriform.

Rami triangular, each with a pair of large alulae; anterior alulae broad, simply rounded or with short point laterally; posterior alulae narrower, acutely pointed, directed posteriorly, incurved; rami tips with 2-3 blunt teeth. Fulcrum tapering posteriorly in ventral view, in lateral view ± parallel-sided in anterior half, thereafter expanding ventrally and narrowing again to rounded end. Unci 5-toothed. Manubria broad; posterior end with short transverse spatulate part ventrally, dorsally with spur-shaped extension; posterior and anterior cavity small. Two epipharyngeal plates, anterior margin bluntly serrate. Two acute pleural rods.

Length 77-85 μm, toe 5-6 μm; trophi 24-25 μm: ramus 10-12 μm, fulcrum 9 μm, uncus 10 μm, manubrium 20-24 μm.

Distribution and ecology
Zaïre (Lake Kivu). Littoral benthos, July, 26.5 °C, pH 7.0.

Species inquirenda

Wulfertia sp.: Dartnall & Hollowday, 1985: 14, figs. 12a-d. (figs. 38,39).
Unci apparently 3-toothed.
Length 130 μm, toe 15 μm; trophi 28 μm.
Antarctica (Signy Island).

3. Genus *Proalinopsis* Weber, 1918

Proalinopsis Weber *in* Weber & Montet (1918), Catalogus Invertébrés Suisse, Genève 11: 98. - Type species: *Proalinopsis caudatus* (Collins, 1873), figs. 40-46.

Diagnosis

Body fusiform, ± slender; illoricate; very transparent. Head and foot separated from trunk by transverse fold. Foot with 1-4 pseudosegments. Toes usually of medium length, acute. Tail region or basal foot pseudosegment with setate or spine bearing dorsal papilla. Dorsal antenna merely a setigerous papilla, or a conspicuous projection. Corona disc-shaped, oblique; buccal field evenly ciliated, circumapical band complete, provided with 2 lateral tufts of long cilia; apical field small. Mouth at or near ventral margin of buccal field. Brain with eyespot. Retrocerebral organ rudimentary or absent.

Trophi virgate or intermediate between malleate and virgate. Rami roughly triangular, in lateral view ± strongly curved at a ± right angle to the fulcrum; basal apophyses well-developed. Fulcrum narrow, short to long. Unci with 4-10 long, slender teeth. Manubria ± long, with lateral lamellae. Epipharynx absent.

Six species.

Cosmopolitan.

Key to species

1 Neck long and slender, without clearly defined posterior limit; posterior dorsal papilla with fine seta(e), at basal pseudosegment of foot (figs. 40-46) .... *1. P. caudatus*
– Head separated from trunk by distinct neck-fold; posterior dorsal papilla with robust spine, at end of trunk 2

2(1) Foot short, < 1/15 total length, one pseudosegment; spine of posterior dorsal papilla spindle-shaped; corona ventral (figs. 47-49) .... **2. *P. selene***
– Foot longer, ≥ 1/7 total length, with 2 pseudosegments; spine of posterior dorsal papilla not spindle-shaped; corona oblique 3

3(2) Foot very long, c. 1/4 total length, terminal pseudosegment wrinkled; posterior dorsal papilla with short, curved spine (figs. 50-53) .... **3. *P. phacus***
– Foot shorter, ≤ 1/7 total length, terminal pseudosegment not wrinkled; posterior dorsal papilla ± long 4

4(3) Toes short, bases strongly swollen; a scaly fold protruding over bases of toes dorsally (figs. 54,55) .... **4. *P. squamipes***
– Toes slender, bases not swollen; without dorsal scaly fold 5

5(4) Foot slender; terminal foot pseudosegment short, c. 1/4 basal one; dorsal antenna a small, setigerous papilla (figs. 56-59) .... **5. *P. gracilis***
– Foot stout; foot pseudosegments ± equally long; dorsal antenna a large knob-like papilla (figs. 60-64) .... **6. *P. staurus***

Descriptions

1. *Proalinopsis caudatus* (Collins, 1873)
Figs. 40-46, pl. 4 figs. 1-6

Synonyms: ? *P. montanus* Godeanu, 1963
? *P. trisegmentus* Sudzuki, 1960

Collins 1873: 11, figs. 8a,b (*Notommata caudata*); Hudson & Gosse 1886: 33, pl. 16 fig. 5 (*Copeus caudatus*); Weber & Montet 1918: 98 (*Proalinopsis caudatus*); Harring & Myers 1922: 608-610, pl. 52 figs. 1-5; Wulfert 1956: 484-485, figs. 38a-f; Koch-Althaus 1963: 432, figs 43a,b; Kutikova 1970: 484, fig. 693; Koste 1978: 267, pl. 87 figs. 4a-f, pl. 88 figs. 3a-m; Koste & Shiel 1990: 131, figs. 1:3a-d.

Type locality
Pools, Sandhurst, Berkshire, England.

Description
Body fusiform, slender; cuticle delicate, outline fairly constant; colourless, hyaline. Adults often with thready gelatinous hull. Head offset by transverse fold, small; neck

long, slender, as wide as head, merging into oval trunk. Trunk flattened ventrally, gibbous dorsally; tail short, collar-shaped. Foot relatively long, c. 1/5-1/7 total length; 2-4 pseudosegments, merging into trunk; basal pseudosegment with cylindrical papilla dorsally, bearing long spine-like seta(e) and collar of small cilia; terminal pseudosegment with small, dorsal knob between toes. Toes moderately long, lanceolate, acute, slightly decurved; basal part with articulating joint. Corona oblique. Dorsal antenna on knob-like or cylindrical papilla. Brain large, saccate. Eyespot bright-red, at posterior of brain. Stomach and intestine separated by constriction. Oesophagus long, slender, cross-lined. Gastric glands often cellular, usually smooth, medium large, elongate-pyriform. Pedal glands elongate, extending into basal foot pseudosegment.

Trophi virgate, symmetrical. Rami triangular; apically near inner margins with 5 large and 5-6 smaller, blunt triangular teeth, interlocking with uncinal teeth; basal apophyses crest-shaped, margins crenate; dorsal and basal fenestrae relatively small, the dorsal laterally in widest part of rami. Fulcrum relatively small, rod-shaped in ventral view, in lateral view knife-shaped. Unci with 7-8 large, clubbed teeth, gradually decreasing in size, and group of 3-4 small teeth near dorsal edge; principal teeth with small projection medially on ventral margin. Manubria broad, with small posterior and long, broad anterior cavity almost extending till posterior, slightly curved, wide lamellar end.

Subitaneous egg elongate oval (fig. 46).

Length 125-268 μm, width up to 77 μm, smallest neck-width 42 μm, toe 12-22 μm; trophi 18 μm: fulcrum 7 μm, uncus 10-11 μm, manubrium 11-18 μm; subitaneous egg 60x30 μm.

Distribution and ecology

Probably cosmopolitan. In *Sphagnum* bogs, periphytic and epibenthic in acid waters, pH 4.5-6.5.

2. *Proalinopsis selene* Myers, 1933

Figs. 47-49

Myers 1933b: 14-16, figs. 9a-c.

Type locality

Witch Hole and Lower Breakneck Pond, Mount Desert Island, Maine, U.S.A.

Description

Body fusiform, slender; cuticle very flexible, outline variable; very hyaline. Head offset by shallow neck-fold, somewhat longer than broad. Trunk cylindrical, posteriorly tapering towards foot. Tail with small, knob-like papilla bearing short, stiff, ± spindle-shaped spine. Foot short, c. 1/18 total length, stout, one pseudosegment. Toes medium long, tapering to slender, acute tips. Corona almost ventral. Dorsal antenna a small setigerous papilla. Brain long. Eyespot small, at posterior of brain. Oesophagus short. Gastric glands medium sized, oval. Pedal glands small, club-shaped.

Trophi intermediate between malleate and virgate, symmetrical. Rami triangular,

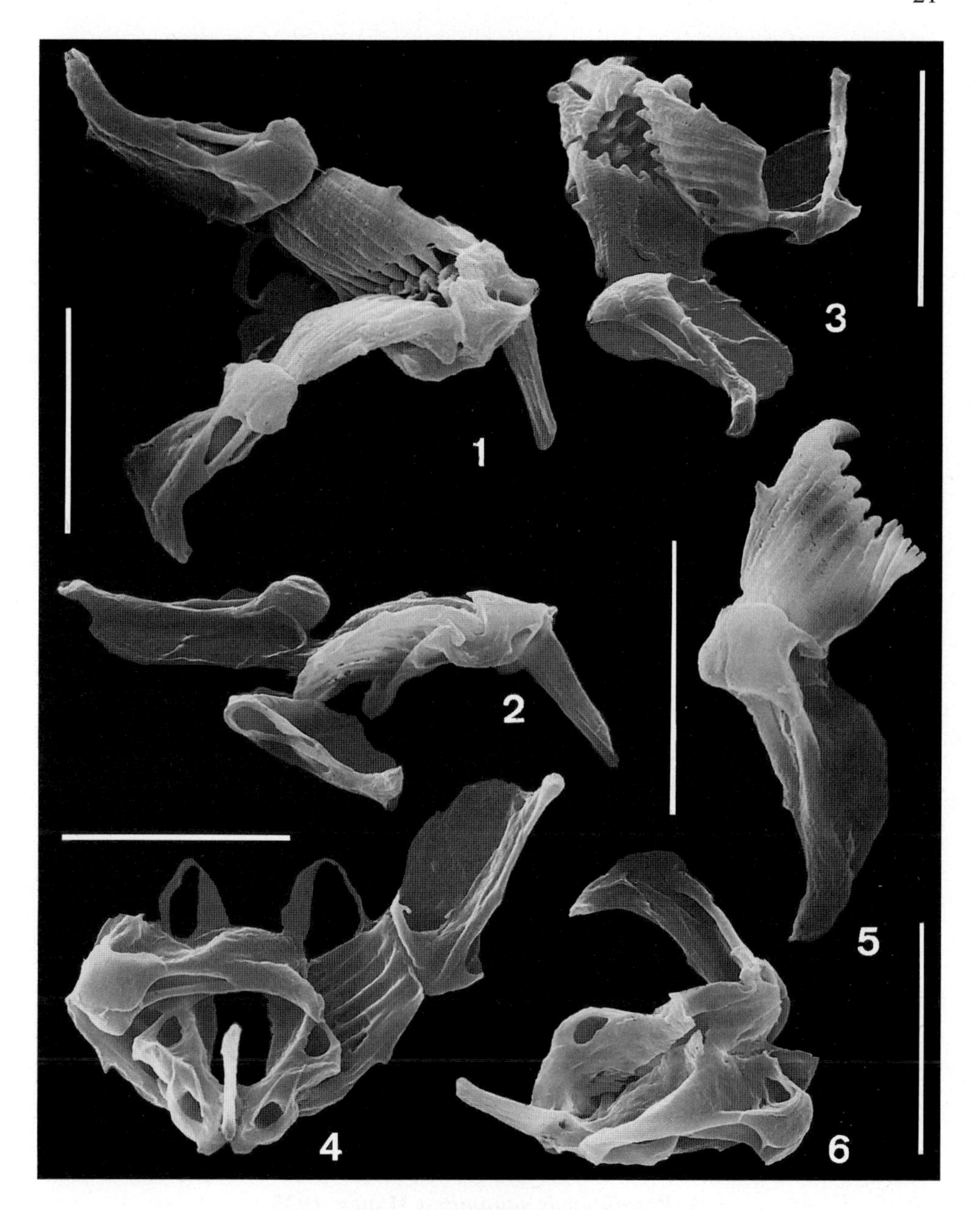

Plate 4. *Proalinopsis caudatus* , trophi (S.E.M. photographs). 1,2: lateral view, 3: apical view, 4: ventro-caudal view, 5: malleus, lateral view, 6: dorso-caudal view.
Scale bars: 10μm.
(1,2: Nome, Alaska; 3-6: Canigou, Pyrenées Orientales, France)

inner margins with c. 8 blunt teeth. Fulcrum long, laterally knife-shaped. Unci with 8 slender teeth, gradually decreasing in size. Manubria short, stout; posterior lamella small, confined to upper half, anterior one large, extending over whole length of manubrium.

Length 90 μm, toe 15 μm.

Distribution and ecology

N. America (Maine). Among algae in acid waters.

3. *Proalinopsis phacus* Myers, 1933

Figs. 50-53

Myers 1933b: 13-14, figs. 8a-d.

Type locality

No single locality specified; Witch Hole, Mount Desert Island, Lenapi Lake, Atlantic County, New Jersey, U.S.A.

Description

Body tapering towards toes, very slender; cuticle very flexible, outline variable. Head offset by shallow neck-fold. Tail with prominent, knob-like papilla bearing very short, curved spine. Foot very long, c. 1/4 total length, slender, 2 pseudosegments of equal length, terminal one wrinkled transversely. Toes medium long, bulbous at base, tips very slender, drawn-out. Corona oblique. Dorsal antenna large, knob-like. Brain short, stout. Eyespot very small, round, at posterior of brain. Oesophagus long, slender. Stomach and intestine separated by shallow constriction. Gastric glands small, kidney-shaped. Pedal glands very long, foot length.

Trophi modified virgate, asymmetrical. Rami triangular, left ramus larger than right, inner margins curved, without teeth; alulae acute. Fulcrum long, rod-shaped in ventral view, laterally knife-shaped. Unci with 4 long, slender teeth, gradually decreasing in size. Left manubrium long, posterior end incurved, right manubrium shorter, almost straight.

Length 165-175 μm, toe 15 μm.

Distribution and ecology

N. America (Maine, New Jersey, Pennsylvania). Among *Sphagnum* in acid waters.

4. *Proalinopsis squamipes* Hauer, 1935

Figs. 54,55

Hauer 1935: 99-100, figs. 26a,b; Kutikova 1970: 485, figs. 692a,b; Koste 1978: 266, pl. 87 figs. 11a,b.

Type locality

Ahabach, Schluchseemoor, Schwarzwald, Germany.

Description
Body fusiform, relatively stout; cuticle flexible, outline variable; fairly hyaline. Head offset by distinct neck-fold. Trunk ± strongly arched dorsally, ventrally flattened. Tail with knob-like papilla bearing stout, straight spine. Foot relatively long, c. 1/8 total length, stout, almost cylindrical; 2 weakly defined pseudosegments. Toes medium long, bases swollen, acute, tips decurved ventrally; a scaly fold on top of base of toes; toes carried closely appressed. Brain large, saccate. Eyespot large, at underside of brain. Oesophagus short, slender.

Trophi with malleate and virgate affinities, symmetrical. Rami triangular; alulae small; basal plates large, triangular, with tooth-like projection at inner edges. Unci with 8-9 teeth, gradually decreasing in size; principal tooth bifurcate. Manubria with lateral lamellae, anterior one almost extending till posterior end.

Length 130 μm, toe 14 μm, spine c. 12 μm.

Distribution and ecology
Europe (Germany, England), N. America (New York, New Jersey, Pennsylvania). In floating *Sphagnum* and aquatic bryophytes, brook, creek, stream; October.

5. *Proalinopsis gracilis* Myers, 1933
Figs. 56-59

Myers 1933b: 11-13, figs. 7a-d.; Godske Eriksen 1969: 21, figs. 10D,E.

Type locality
No single locality specified; Mount Desert Island; Atlantic County, New Jersey; Villas County, Wisconsin; Montgomery County, Pennsylvania, U.S.A.

Description
Body slender, tapering towards toes; cuticle flexible, outline fairly constant. Head offset by distinct neck-fold, somewhat longer than wide. Tail with small, knob-like papilla bearing long, stiff spine. Foot relatively long, c. 1/8 total length, slender, 2 pseudosegments, terminal one 1/4 length of basal. Toes medium long, tips drawn out, acute. Corona oblique. Dorsal antenna a small setigerous papilla. Brain saccate, a small, round eyespot somewhat ventrally to its posterior. Oesophagus long, slender. No constriction between stomach and intestine. Gastric glands small, oval. Pedal glands elongate, foot length.

Trophi virgate, slightly asymmetrical. Rami triangular, inner margins without teeth; alulae prominent, left much longer than right. Fulcrum moderately long, rod-shaped in ventral view, laterally knife-shaped. Unci with one main tooth and 4 subsidiary teeth. Manubria asymmetrical, left larger than right, posterior end curved inwardly; anterior and posterior lamellae < 1/2 manubrium length.

Length 133-140 μm, toe 15-18 μm.

Distribution and ecology
N. America (New Jersey, Wisconsin, Pennsylvania, Maine), Europe (Sweden, Finland). Among vegetation in ponds.

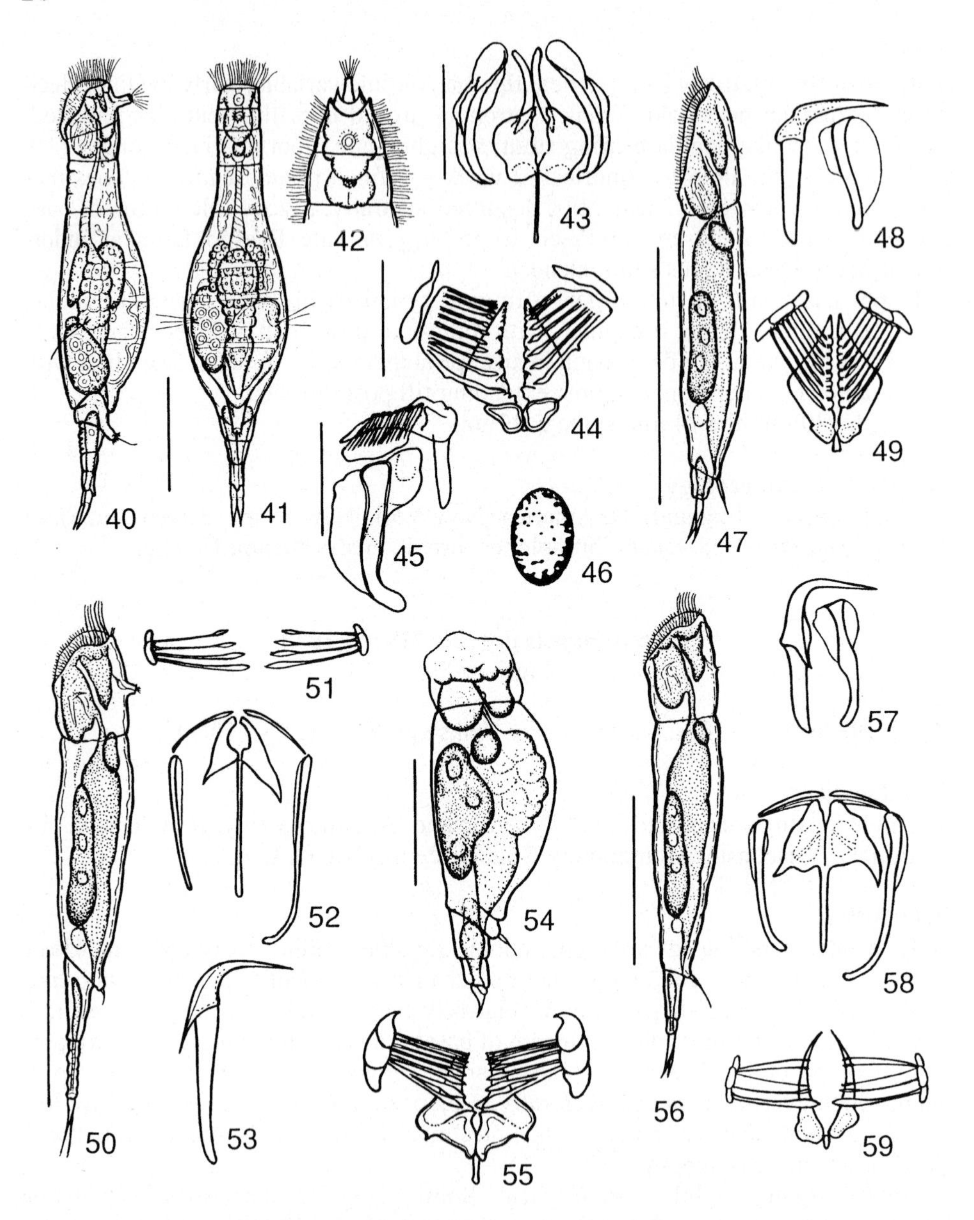

Figs. 40-46: *Proalinopsis caudatus*. 40: lateral view, 41: dorsal view, 42: head, detail, 43: trophi, dorsal view, 44: trophi, apical view, 45: trophi, lateral view, 46: subitaneous egg.
Figs. 47-49: *Proalinopsis selene*. 47: lateral view, 48: incus and manubrium, lateral view, 49: trophi, apical view.
Figs. 50-53: *Proalinopsis phacus*. 50: lateral view, 51: unci, 52: trophi, ventral view, 53: incus, lateral view.
Figs. 54,55: *Proalinopsis squamipes*. 54: lateral view, 55: trophi, apical view.
Figs. 56-59. *Proalinopsis gracilis*. 56: lateral view, 57: incus and manubrium, lateral view, 58: trophi, ventral view, 59: trophi, apical view.
Scale bars: habitus 50 µm, trophi 10 µm.
(40-42,46: after Wulfert, 1956; 43-45: Canigou, Pyrenées Orientales, France; 47-53,56-59: after Myers, 1933b; 54,55: after Hauer, 1935)

6. *Proalinopsis staurus* Harring & Myers, 1924
Figs. 60-64

Harring & Myers 1924: 439-440, pl. 20 figs. 5-9; Remane 1929-33: 543, figs. 322a-e; Koste 1978: 266, pl. 87 figs. 5a,b; Koste & Shiel 1990: 131, fig. 1:4.

Type locality
No single locality specified; Mamie Lake, Eagle River and Lac Vieux Desert, Villas County, Wisconsin; also New Jersey and Florida, U.S.A.

Description
Body fusiform, fairly slender; cuticle flexible, outline variable; hyaline. Head offset by distinct neck-fold, subquadratic. Trunk widest near middle. Tail with knob-like papilla bearing long, c. 1/5 total length, stiff spine. Foot relatively long, c. 1/7 total length, stout, 2 pseudosegments, terminal one somewhat longer than basal. Toes ± long, rather stout at base, gradually tapering to drawn out, acute tips. Corona oblique. Dorsal antenna knob-like. Brain large, saccate. Retrocerebral organ absent. Eyespot cervical, sometimes very pale and hard to see. Oesophagus long, slender. No constriction between stomach and intestine. Gastric glands small, spherical. Pedal glands large, pyriform, foot length.

Trophi intermediate between malleate and virgate, symmetrical. Rami triangular, apparently without teeth at inner margins; alulae small; basal apophyses large, triangular. Fulcrum short, rod-shaped in ventral view, laterally with broad base, tapering towards slightly expanded, scalloped posterior end. Unci with 8-9 teeth, gradually decreasing in size, principal tooth bifurcate. Manubria very long, posterior lamella small, anterior one almost extending till posterior end.

Length 100 μm, toe 18 μm, spine 22 μm; trophi 15 μm.

Distribution and ecology
N. America (Wisconsin, New Jersey, Pennsylvania, Florida), S. America (Matto Grosso, Brazil), Australia (Tasmania), ?Africa (Cameroon). Among submerged *Sphagnum*, lakes and ponds; 18.0-25.0 °C, pH 5.8-7.8, 40.8-46.6 μScm^{-1}.

Species inquirendae

Proalinopsis lobatus: Rodewald, 1935: 205-207, figs. 7a,b, 8a-c. (figs. 65-69).
Body fusiform. Cuticle flexible, outline variable. Head offset by neck-fold. Trunk with 3 pseudosegments, separated by broad transverse folds. Tail trilobate, a long spine inserted on central lobe. Foot with 2 pseudosegments. Toes short, acutely pointed. Eyespot elongate, on brain. Retrocerebral organ absent. Dorsal antenna long, cylindrical.

Trophi virgate. Rami triangular; basal apophyses large. Fulcrum short, posterior end expanded. Unci 7-toothed.

Length 163-193 μm, width 42-50 μm, foot 28-35 μm, toe 18-20 μm, caudal spine 35 μm.

Europe (Romania, Slovakia). In montane *Sphagnum* bog; April-June.

Contraction form ? Description of trophi unsatisfactory.

Proalinopsis montanus: Godeanu, 1963: 374-376, figs. 1a-f. (figs. 70-72).
General outline of body similar to *P. caudatus*. Head broader than neck. Corona with one median and 2 lateral tufts of long cilia. Base of toes inflated, with shallow constriction. Trophi insufficiently described.

Length 210 µm, width 51 µm, spine 15 µm, toe 12 µm; trophi 21 µm.

Europe (Romania). In littoral plankton from small permanent pond without vegetation, in mountain streams; June, 10.5 °C, pH 5.2.

The description is considered insufficient, to permit a distinction to be made between *P. caudatus* and *P. montanus*.

Proalinopsis trisegmentus: Sudzuki, 1960: 24-25, figs. 5a-e.
Foot with 3 pseudosegments. Dorso-caudal papilla at end of trunk, double knobbed, with a tuft of short setae. Trophi insufficiently described, not figured. Probably synonymous with *P. caudatus*.

4. Genus *Proales* Gosse, 1886

Proales Gosse *in* Hudson & Gosse (1886), Rotifera 2: 36. - Type species: *Proales decipiens* (Ehrenberg, 1832), figs. 275-286.

Diagnosis

Body fusiform, vasiform, roughly cylindrical, or tapering towards toes. Usually illoricate, rarely semiloricate. Head and trunk separated by neck-fold. Trunk usually tapering towards base of foot. Foot varying greatly in length and shape with different species: indistinct or short, medium long to very long, pseudosegments present or absent. Two ± short toes (except *P. doliaris* with one toe). Corona usually an oblique disc, with well-developed marginal cilia and 2 lateral tufts of densely set, long cilia; without retractile ciliated auricles. Buccal field large, evenly ciliated, with mouth at ventral margin. Apical field not usually enclosed by marginal ciliation, sometimes dorsal. Rostrum slightly developed in some species. Eyespot usually on or near brain, rarely frontal or absent. Retrocerebral organ present, rudimentary or absent.

Mastax often very small; trophi malleate, a modification of the malleate type or virgate. Rami roughly triangular, usually with teeth at their inner margins, basal apophyses large. Fulcrum usually wedge-shaped, short to medium long, ± in line with the flattened rami. Unci with 1-7 teeth. Manubria ± long. Epipharynx two rod- or irregular, wedge-shaped pieces.

About 43 species.

Cosmopolitan.

Free-living, epiphytic, epizoic or parasitic. Freshwater to halophile.

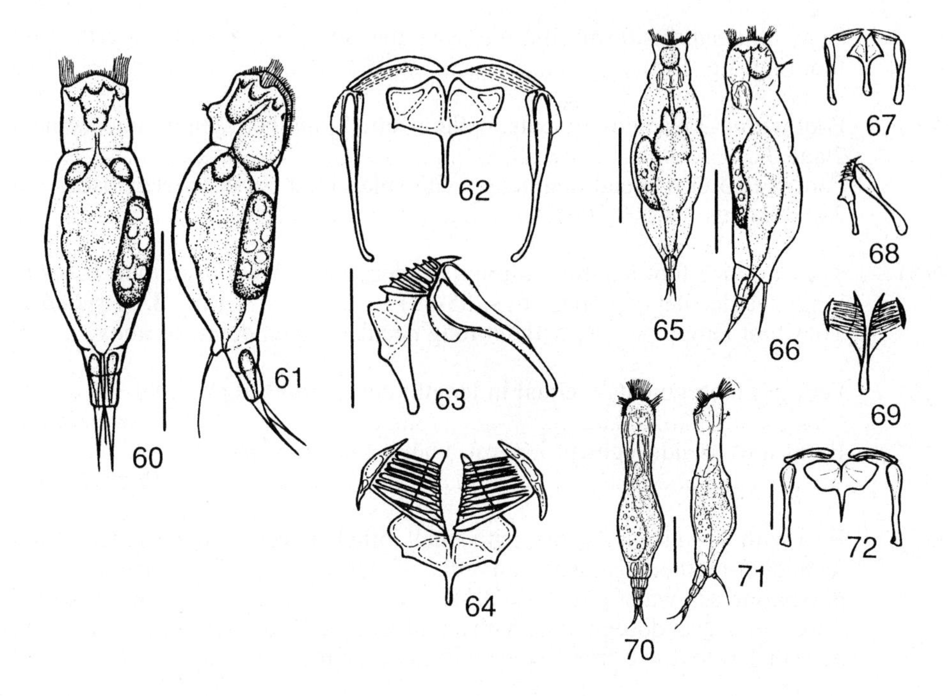

Figs. 60-64: *Proalinopsis staurus*. 60: dorsal view, 61: lateral view, 62: trophi, ventral view, 63: trophi, lateral view, 64: trophi, apical view.
Figs. 65-69: *Proalinopsis lobatus*. 65: dorsal view, 66: lateral view, 67: trophi, ventral view, 68: trophi, lateral view, 69: trophi, apical view.
Figs. 70-72: *Proalinopsis montanus*. 70: dorsal view, 71: lateral view, 72: trophi, dorsal view.
Scale bars: habitus 50 μm, trophi 10 μm.
(60-64: after Harring & Myers, 1924; 65-69: after Rodewald, 1935; 70-72: after Godeanu, 1963)

Key to species

1 In brackish, marine or inland saline water 2
– In freshwater 12

2(1) Corona ventral; epizoic or ectoparasitic on hydroid polyps 3
– Corona frontal to oblique; not epizoic or ectoparasitic on hydroid polyps 4

3(2) Tail absent; subcerebral glands absent; rami without denticulate basal apophyses; with dome-shaped epipharynx (figs. 73-77) .... **1. *P. gonothyraeae***
– With posteriorly rounded tail; subcerebral glands present; rami with denticulate basal apophyses; dome-shaped epipharynx absent (figs. 78-84) .... **2. *P. christinae***

4(2) Body vasiform in dorsal view; foot inclusive toes long, abruptly offset from trunk (e.g. fig. 95) 5

– Body elongate fusiform; foot inclusive toes long, ± gradually tapering into foot (e.g. fig. 135) 10

5(4) Foot with 2-3 pseudosegments, without small joint between pseudosegment 2 and 3 6
– Foot with 3 or 4 pseudosegments, with small joint between pseudosegment 2 and 3 (figs. 98,102,106) 8

6(5) Toes c. twice foot length, ± abruptly reduced at 1/3-1/2 from their bases to long, needle-shaped points (figs. 85-89) .... **3. *P. syltensis***
– Toes foot length or less, without long needle-shaped points as above 7

7(6) Foot pseudosegments ± equal in length; unci 7-toothed (figs. 90-93) .... **4. *P. commutata***
– Basal foot pseudosegment longest; unci 4-toothed (figs. 94-97) .... **5. *P. germanica***

8(5) Foot with 4 pseudosegments, with small joint between pseudosegment 2 and 3; unci 5-toothed; parasitic on gills of hermit crab (*Eupagurus*); meso-polyhaline sea-water (figs. 98-101) .... **6. *P. paguri***
– Foot with 3 pseudosegments, with small joint between pseudosegment 2 and 3; unci 1-6-toothed; free-living; oligo-polyhaline sea-water 9

9(8) Unci with single principal tooth and 1/2 accessory toothlets; mesohaline sea-water (figs. 102-105) .... **7. *P. oculata***
– Unci 5-6-toothed; oligo-polyhaline sea water (figs. 106-110) .... **8. *P. halophila***

10(4) Eyespot at posterior of brain; foot one pseudosegment, wrinkled (figs. 116-128) .... **9. *P. similis***
– Eyespot(s) frontal (occasionally absent in adults); foot with 2-3 pseudo-segments 11

11 (10) Foot with 3 pseudosegments, terminal pseudosegment shortest; toes partially fused at their bases; posterior end of fulcrum strongly expanded; fresh-, brackish- and marine water (figs. 129-134) .... **10. *P. theodora***
– Foot with 2 pseudosegments, basal pseudosegment shortest; toes not fused at their bases; posterior end of fulcrum not strongly expanded; oligo-polyhaline sea-water (figs. 135-141) .... **11. *P. reinhardti***

12(1) Eyespot(s) frontal (occasionally absent in adults) 13
– Eyespot(s) cerebral, at posterior, below base or on top of brain, or absent 15

13(12) Foot long, ≥ 1/3 total length, 3 pseudosegments, terminal pseudosegment very short, usually partially retracted in median pseudosegment (figs. 129-134) .... **10. *P. theodora***
– Foot shorter, 1/4-1/5 total length, 2-3 pseudosegments, terminal pseudo-segment longest 14

14(13) Four paired accumulations of light-refracting granules near gastric glands and protonephridia (figs. 142-148) .... **12. *P. globulifera***
– Paired accumulations of light-refracting granules absent (figs. 149-153) ... . **13. *P. minima***

15(12) Foot with single toe (figs. 154-160) .... **14. *P. doliaris***
– Foot with 2 toes 16

16(15) Tail prominent, covering foot completely; body cylindrical 17
– Tail less prominent, minute or only lying over basal foot pseudosegment; body tapering towards toes, fusiform, roughly cylindrical or saccate 18

17(16) Toes short, c. 1/34 total length, conical; gastric glands elongated with acute anterior ends; stomach and intestine separated by constriction; rostrum small (figs. 161-164) .... **15. *P. cryptopus***
– Toes longer, c. 1/17 total length, ± lanceolate; gastric glands kidney-shaped; stomach and intestine not separated by constriction; rostrum prominent (figs. 165-169) .... **16. *P. macrura***

18(16) Foot with 4 pseudosegments 19
– Foot with 1-3 pseudosegments 20

19(18) Foot stout, terminal pseudosegment broader than long; bases of toes partially fused; retrocerebral sac large, hemispherical, at posterior of brain (figs. 170-176) .... **17. *P. alba***
– Foot slender, terminal pseudosegment longer than broad; bases of toes not fused; retrocerebral sac small, confluent on dorsal side of brain (figs. 177-179) .... **18. *P. bemata***

20(18) Foot one pseudosegment 21
– Foot with 2-3 pseudosegments 30

21(20) Foot distinctly offset from trunk (figs. 180,184) 22
– Foot continuous with general outline of body 23

22(21) Toes finger-shaped, relatively long, ratio foot:toe = 1:2; unci 6-toothed (figs. 180-183) .... **19. *P.indirae***
– Toes short, conical, ratio foot:toe ≥ 2.0; unci 4-toothed (figs. 184-189)...... .... **20. *P. baradlana***

23(21) Without dorsal papilla or spur between toes 24
– With dorsal papilla or spur between toes 28

24(23) Trunk with 7 strong, regularly placed transverse folds; 2 large eyespots, symmetrically to median line (figs. 190-193) .... **21. *P. provida***
– Trunk without such strong transverse folds; a single, small eyespot, or 2 eyespots, the main one displaced to right, the secondary one ± median 25

25(24) Corona nearly ventral; toes slender, c. 1/10 total length, acute; stomach and

intestine separated by deep constriction; trophi asymmetrical (figs. 194-196) .... **22. *P. gladia***
– Corona nearly frontal or oblique; toes short, ≤ 1/13 total length, conical; stomach not sharply marked-off from intestine; trophi symmetrical........ 26

26(25) Body tapering towards toes, head as wide as trunk, with rostrum-like prominence; unci single-toothed with serrate lamellar plate (figs. 203-205) .... **23. *P. simplex***
– Body fusiform, head narrower than trunk, without rostrum-like prominence; unci 4-toothed without serrate lamellar plate 27

27(26) Body fusiform, stout, swollen; eyespot minute, at posterior of brain; pedal glands large, pyriform; with glandular caecum (figs. 197-202) .... **24. *P. segnis***
– Body elongate fusiform, slender; 2 eyespots posteriorly on brain, main one displaced to right , secondary one ± median; pedal glands slender, elongate; without glandular caecum (figs. 206-209) .... **25. *P. phaeopis***

28(23) Toes ± long, ≥ 1/8 total length, dagger-like; corona ± frontal; stomach and intestine separated by constriction; eyespot at posterior of brain, median (figs. 210-214) .... **26. *P. pugio***
– Toes short, c. 1/14-1/18 total length, conical; corona oblique; stomach and intestine not separated by constriction; eyespot rudimentary, on brain, displaced to right 29

29(28) Body slender, trunk abruptly reduced at half-length, continuing ± parallel-sided; a dorsally projecting spur between toes; pedal glands foot length, without prominent mucus reservoir; rami laterally each with anteriorly projecting process at 1/3 from tip (figs. 215-219) .... **27. *P. palimmeka***
– Body stout, trunk gradually tapering posteriorly; a bulbous process surmounted by papilla dorsally between toes; pedal glands extending into trunk, with prominent mucus reservoirs; outer margins of rami without lateral process (figs. 220-224) .... **28. *P. cognita***

30(20) With dorsal papilla or spur between toes 31
– Without dorsal papilla or spur between toes 33

31(30) With pointed spur, dorsally between toes; body plump fusiform, cylindrical; free-swimming and parasitic in eggs of water snails (figs. 225-233) .... **29. *P. gigantea***
– With small, rounded dorsal papilla between toes; body slender, elongate fusiform, cylindrical; free-living 32

32(31) Dorsal papilla between toes simple; dorsal margin of terminal foot pseudosegment without lateral spurs; basal foot pseudosegment twice length terminal one; stomach and intestine separated by constriction (figs. 234-243) .... **30. *P. fallaciosa***
– A dorsal bulbous process surmounted by papilla between toes; dorsal margin of terminal foot pseudosegment with lateral spurs; terminal foot pseudosegment twice length basal one; stomach and intestine not separatedby constriction (figs. 244-248) .... **31. *P. ornata***

33(30) Basal foot pseudosegment completely covered by tail; eyespot displaced somewhat to left; greatest body width in posterior 1/4 (figs. 249-252) .... **32. *P. wesenbergi***

– Basal foot pseudosegment not covered by tail; eyespot(s) median or displaced to right; greatest body width in anterior 3/4 34

34(33) Body fusiform, ± slender, elongate, cylindrical; ± slightly curved dorsally in lateral view 35

– Body fusiform, stout; ± strongly arched dorsally in lateral view 41

35(34) Foot with 3 pseudosegments 36

– Foot with 2 pseudosegments 37

36(35) Foot long, c. 1/4 total length; terminal foot pseudosegment projecting over base of toes; unci 5-toothed (figs.253-259) .... **33. *P. sordida***

– Foot short, ≤ 1/6 total length; terminal foot pseudosegment not projecting over base of toes; unci 3-toothed (figs. 206-266) .... **34. *P. micropus***

37(35) Foot c. 1/5 total length; terminal foot pseudosegment twice length basal one (figs. 267-270) .... **35. *P. adenodis***

– Foot ≤ 1/7 total length; terminal foot pseudosegment shorter or equal in length to basal one 38

38(37) Free-living 39

– Parasitic in filamentous and colonial algae or colonial ciliates 40

39(38) Terminal foot pseudosegment with incision between toes; retrocerebral sac rudimentary; unci 2-toothed (figs. 271-274) .... **36. *P. granulosa***

– Terminal foot pseudosegment without incision between toes; retrocerebral sac distinct, pyriform; unci with 4/5 or 5/5 teeth (figs. 275-286) .... **37. *P. decipiens***

40(38) Parasitic in filaments of siphonous algae (*Vaucheria*, *Dichotomosiphon*), gall-forming; unci single-toothed (figs. 287-295) .... **38. *P. werneckii***

– Parasitic in colonial algae (*Volvox*, *Uroglena*, *Uroglenopsis*) and in colonies of ciliates (*Ophrydium*); unci 3-toothed (figs. 296-306) .... **39. *P. parasita***

41(34) Foot with 3 pseudosegments (sometimes indistinct) 42

– Foot with 2 pseudosegments 43

42(41) Toes short, cylindrical, with ± abruptly reduced, blunt tips; eyespot absent; rami outline quadratic (figs. 307-309) .... **40. *P. lenta***

– Toes short, conical in lateral view, foliate and ± acutely pointed in dorsal view; eyespot at posterior of brain; rami outline triangular (figs. 310-318) .... **41. *P. sigmoidea***

43(41) Basal apophyses large horn-like projections (figs. 330-336) .... **42. *P. daphnicola***

– Basal apophyses large, spoon-shaped (figs. 319-329) .... **43. *P. kostei***

Descriptions

1. *Proales gonothyraeae* Remane, 1929
Figs. 73-77

Remane 1929a: 289-295, figs. 1-3; Remane 1929b: 64-66, figs. 43-45; Remane 1929-33: 540-542; Voigt 1957: 243, figs. 1, 8a,b; Kutikova 1970: 496, figs. 715a-e; Koste 1978: 271-272, figs. 14.4, 26.1, 48.3, pl. 91 figs. 7a-7e.

Type locality
Kieler Bucht, Baltic Sea, Germany; in thecae of hydroid polyp *Laomedea loveni* (Allmann).

Description
Body fusiform, widest in posterior third; cuticle thin, flexible. Animals within theca of hydroids have their foot retracted and outline variable. Head and neck offset by shallow, dorsal transverse folds; head sometimes with additional transverse fold, rounded anteriorly. Foot moderately short, c. 1/5 total length, 2 pseudosegments. Toes short, broad, conical. Corona nearly ventral, pre-oral cilia in 7 transverse rows, laterally 2 tentacles, lateral ciliary tufts absent. Brain egg-shaped, surrounded by large retrocerebral sac; 2 ducts with light-refracting bodies; in sac 6 nuclei with large nucleoli. Two red, close-set, frontal eyes without lenses. No constriction between stomach and intestine. Gastric glands medium large, globular, slightly stalked. Pedal glands large, elongated, extending into trunk, with small reservoir in toes.

Trophi malleovirgate. Incus small. Rami triangular, anteriorly pointed; alulae present; no basal apophyses. Fulcrum short, in ventral view rod-shaped, posterior end disc-shaped; laterally keeled. Unci fairly thin plates with 4-5 ribs. Manubria long, slender, incurved posteriorly, head with 2 small lamellae. Epipharynx large, dome-shaped; 2 club-shaped oral plates.

Length 250-320 μm, foot 70 μm, toe 20 μm; incus 20 μm, ramus with alula 15 μm, fulcrum 11 μm, uncus 9 μm, manubrium 25 μm.

Distribution and ecology
Baltic Sea (Germany), ? Channel (France). In thecae of hydroid polyp *Laomedea* (syn. *Gonothyraea*) *loveni* (Allmann); ectoparasitic; subitaneous eggs attached to inner and outer side of hydroid theca.

2. *Proales christinae* De Smet, 1994
Figs. 78-84, pl. 5 figs. 1-4

De Smet 1994: 21-25, figs. 1-9.

Type locality and types
Westende, beach of North Sea, Belgium; among hydroids. Holotype and paratypes in the Koninklijk Belgisch Instituut voor Natuurwetenschappen, Brussels, Belgium, No. AI 28.040; paratypes in R.U.C.A.

Description

Body elongate, stout, fusiform in dorsal view, ventrally bent in lateral view; broadly oval in cross-section, higher than wide, greatest diameter near mid-length, fairly hyaline. Head and neck region narrower than trunk; head offset by dorsal transverse fold, occasionally with dorsal transverse fold anteriorly; neck offset from trunk by shallow, dorsal transverse fold. Trunk arched dorsally, more or less abruptly narrowing posteriorly; tail distinct, rounded posteriorly. Foot moderately short, c. 1/5 total length, two pseudosegments of equal length. Toes straight, more or less lanceolate in lateral view, abruptly ending in tubular points; in dorsal view with straight inner margins and curved outer margins; inner margins with short indentation prior to tubules, tubules laterally outcurved prior to their free end. Corona ventral, an anterior row of close-set cilia (reduced circumapical band ?) and at some distance (reduced apical field?) from the latter six tansverse, pre-oral bands of close-set cilia; lateral ciliary tufts absent. Dorsal antenna unpaired, short. Mouth near posterior edge of corona. Retrocerebral organ with sac and a pair of subcerebral glands. Brain egg-shaped, surrounded by large retrocerebral sac, two ducts, sac and ducts granular. Eyespots absent ? No distinct constriction between stomach and intestine. Gastric glands medium large, globular, slightly compressed laterally, short-stalked. Pedal glands large, elongated, extending into trunk, with reservoir in toes. Vitellarium rounded, 8 nuclei.

Trophi modified malleate. Rami triangular, inner margins, smooth, three to four short projections prior to tip; basal apophyses asymmetric: right apophyse small, 3-toothed, left large, triangular, 7-toothed. Fulcrum short, in ventral view rod-shaped, slightly expanded posteriorly; in lateral view with expanded, ventrally recurved, hook-shaped posterior end. Left and right uncus with one principal and 5-6 subsidiary teeth, principal teeth with single accessory toothlet; principal teeth slightly clubbed, others linear, a supplementary comb of sharp, delicate teeth underneath tips of uncinal teeth. Manubria long, slender, strongly incurved posteriorly, head with short lateral cavities, anterior broadest . Two small, club-shaped epipharyngeal elements.

Subitaneous egg elongate oval in dorsal view, ventrally flattened in lateral view; short-stalked; smooth.

Length 270-320 μm, heigth 74-88 μm, width 65-70 μm, toe 32-36 μm; trophi: ramus 12-14 μm, fulcrum 6-7 μm, uncus 12-15 μm, manubrium 16-20 μm; subitaneous egg (LxWxH): 114-122x45-55x45-51 μm.

Distribution and ecology

North Sea, Belgium. Among hydroids washed ashore on the beach.

3. *Proales syltensis* Tzschaschel, 1978

Figs. 85-89

Tzschaschel *in* Koste 1978: 275, pl. 89a figs. 3a-e; Tzschaschel 1979: 10-12, figs. 3a-f.

Type locality

Sandy beach near station of former Biological Institute, Helgoland, List, North Sea, Germany.

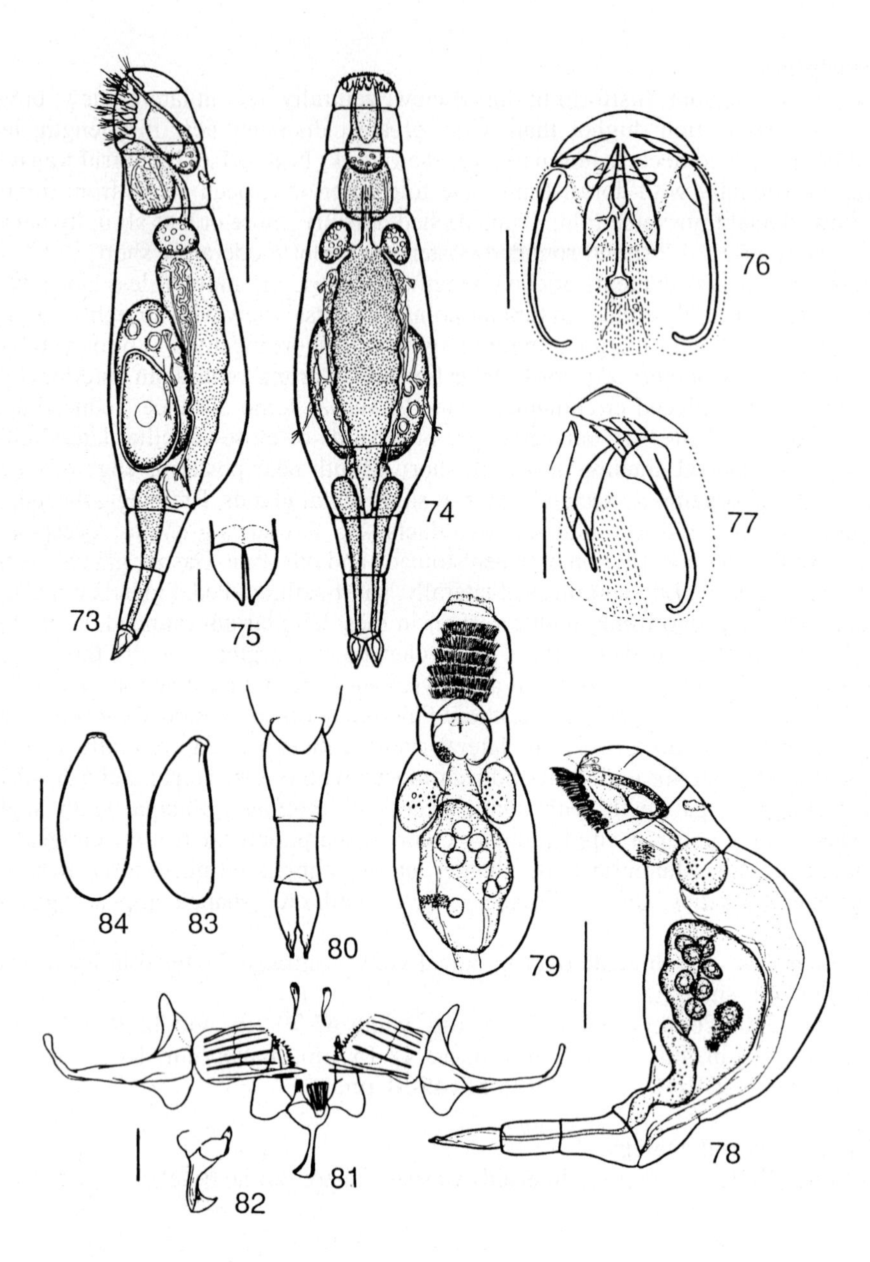

Figs. 73-77: *Proales gonothyraeae*. 73: lateral view, 74: dorsal view, 75: toes, 76: trophi, ventral view, 77: trophi, lateral view.
Figs. 78-84: *Proales christinae*. 78: lateral view, 79: ventral view (foot omitted), 80: detail foot and toes, 81: trophi, ventral view, 82: fulcrum, lateral view, 83: subitaneous egg, lateral view, 84: subitaneous egg, dorsal view.
Scale bars: habitus 50 μm, trophi and detail toes (75) 10 μm, eggs 50 μm.
(73-77: after Remane, 1929a; 78-84: after De Smet, 1994)

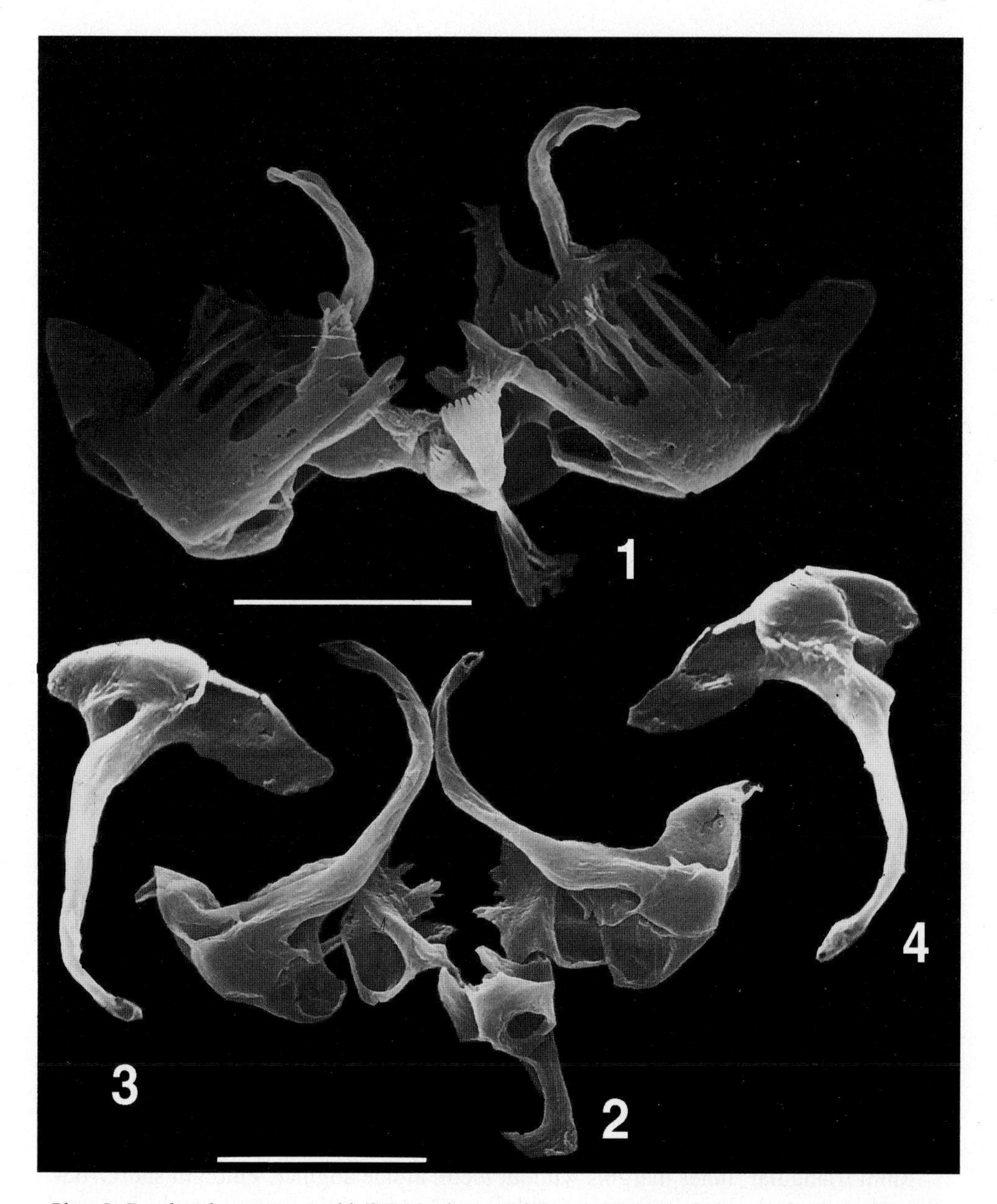

Plate 5. *Proales christinae* , trophi (S.E.M. photographs). 1: ventro-apical view, 2: dorso-caudal view, 3,4: manubria.
(1-4: North Sea, Westende, Belgium)

Description

Body fusiform, ± slender; semiloricate. Head offset by neck-fold, relatively broad, quadratic, with distinct frontal and shallow medial transverse fold. Trunk flattened ventrally, slightly convex dorsally. Foot fairly short, c. 1/9 total length, 3 pseudosegments. Toes fairly long, slender, flexible, ± abruptly reduced to long, needle-shaped points, broad proximal part with small reservoirs. Corona ± frontal, lateral ciliary tufts absent. Brain large, saccate. Retrocerebral sac rudimentary, at end of brain, Y-shaped duct on brain. Eyespot and subcerebral glands absent. Proventriculus present. Stomach wall with 5 thickened rings, inner walls of posterior 2 strongly ciliated. Intestine thin-walled, clearly separated from stomach. Gastric glands kidney-shaped with long stalks and accumulation of small granules anterodorsally. Pedal glands elongated pyriform, in posterior of trunk, long ducts without visible secretion, 2 small club-shaped reservoirs in foot pseudosegment.

Trophi malleate with strong virgate affinities. Rami triangular, with dorsally curved lamella; small triangular alulae; basal apophyses small. Fulcrum in ventral view rod-shaped, short; slightly expanded and curved in lateral view. Left uncus with 5 teeth, gradually decreasing in size. Manubria relatively long, posterior end incurved at a right angle, laterally broad with anterior and posterior lamellae.

Subitaneous eggs small, oval, short-stalked.

Length 150 μm, toe 40 μm; trophi 17 μm: ramus 6 μm, fulcrum 2 μm, uncus 7 μm, manubrium 13 μm.

Distribution and ecology

North Sea (Germany). Littoral psammon; frequent in July-September, sporadic in spring. Literature: Tzschaschel (1980).

4. *Proales commutata* Althaus, 1957

Figs. 90-93

Althaus 1957: 455, figs. 14a-f; Rudescu 1961: 299-300, figs. 4a-f; Kutikova 1970: 490, figs. 701 a-e.

Type locality

Mesopsammon 2 and 3 m, Spatnite pjasatzi, north of Varna, Black Sea, Bulgaria.

Description

Body vasiform in dorsal view. Head short, offset by neck-fold. Trunk broadly elliptical, posterior with semi-circular plate. Foot moderately long, c. 1/5 total length, stout, with 3 pseudosegments of ± equal length. Toes fairly long, slender, slightly decurved ventrally, almost cylindrical half their length, continuing tapering to acute points. Corona slightly oblique. Two frontal eyes, close-set, dark red.

Rami triangular, inner margins without teeth, alulae angular. Fulcrum short, rod-shaped in ventral view. Left uncus with 7 teeth. Manubria long, strongly incurved posteriorly, apparently without lamellae.

Length 118 μm, toe 23-24 μm; trophi 21 μm: fulcrum 5 μm, manubrium 13 μm.

Distribution and ecology

Black Sea (Bulgaria). Mesopsammon, 2 and 3 m; salinity 16-18‰.

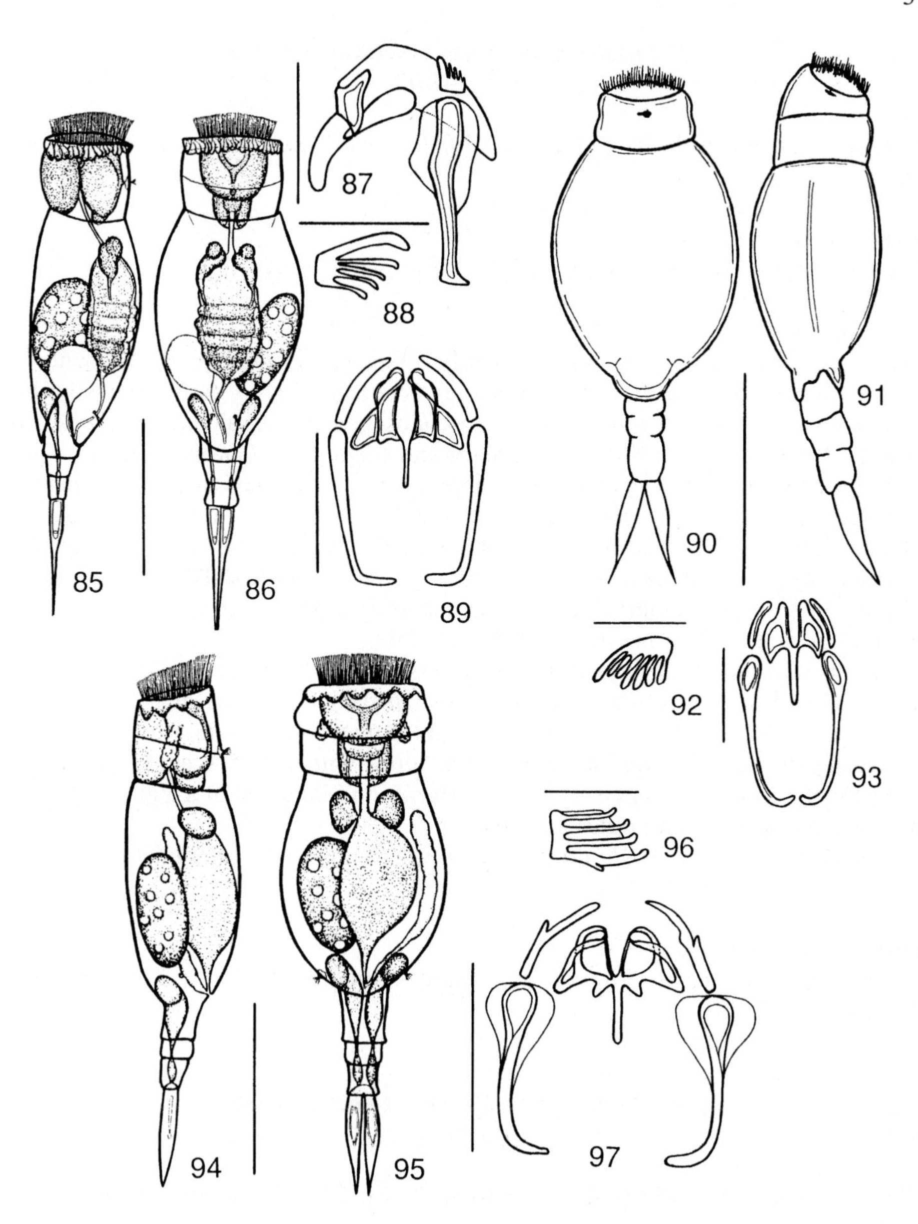

Figs. 85-89: *Proales syltensis*. 85: lateral view, 86: dorsal view, 87: trophi, lateral view, 88: left uncus, 89: trophi, ventral view.
Figs. 90-93: *Proales commutata*. 90: dorsal view, 91: lateral view, 92: left uncus, 93: trophi.
Figs. 94-97: *Proales germanica*. 94: lateral view, 95: dorsal view, 96: left uncus, 97: trophi, ventral view.
Scale bars: habitus 50 μm, trophi 10 μm (92,96: 5 μm).
(85-89,94-97: after Tzschaschel, 1979; 90-93: after Althaus, 1957).

5. *Proales germanica* Tzschaschel, 1978
Figs. 94-97

Tzschaschel *in* Koste 1978: 274-275, pl. 89a figs. 2a-d; Tzschaschel 1979: 8-10, figs. 2a-d;.

Type locality
Shoal east from bird island Uthörn, Sylt, North Sea, Germany.

Description
Body vasiform in dorsal view. Head clearly offset, short, relatively broad, with dorsal transverse fold. Trunk elliptical, flattened ventrally, slightly arched dorsally. Foot moderately long, c. 1/5 total length, 3 pseudosegments, basal longest. Toes long, slender, lanceolate, each with 2 small ducts and single pore. Corona slightly oblique, lateral tufts of long cilia absent. Brain large, saccate; Y-shaped duct of retrocerebral organ on brain; sac posterior to brain, distally coarsely granulated; subcerebral glands wart-like. Eyespot absent. Intestine weakly developed, not separated from stomach by constriction. Gastric glands spherical, short-stalked. Pedal glands distally in trunk, globular; ducts broad, tapering towards posterior of basal foot pseudosegment, ending in club-shaped reservoirs.

Trophi malleate with virgate modifications. Rami small, triangular, latero-distal ends curved downwards, inwardly with rod-shaped suprarami; basal apophyses well-developed. Fulcrum short, rod-shaped in ventral view. Unci with 4 teeth, gradually decreasing in size; left and right principal tooth with 2 and 1 accessory toothlet respectively. Manubria relatively long, posterior end incurved at a right angle, head expanded, with inner and outer lamella.

Length 145 μm, toe 30-35 μm; trophi 15 μm: ramus 4 μm, fulcrum 3 μm, uncus 5 μm, manubrium 9 μm.

Distribution and ecology
North Sea (Germany). Littoral psammon; spring, July-September. Literature: Tzschaschel (1980).

6. *Proales paguri* Thane-Fenchel, 1966
Figs. 98-101

Thane-Fenchel 1966: 95-96, figs. 1-2; Koste 1978: 273-274, pl. 89 figs. 9a-d.

Type locality
No single locality specified, “from hermit crabs taken in dredge hauls at Ellekilde Hageand Knähaken (depth: 25-30 m) in the northern part of the Øresund,... Helgoland”, North Sea, Denmark, Germany.

Description
Body vasiform in dorsal view. Light-refracting bodies dispersed all over trunk and head. Head short, offset by neck-fold. Trunk elliptical. Foot long, c. 1/4 total length, stout, with 5 pseudosegments, the midst shortest; retractable in trunk. Toes long,

broadly lanceolate. Corona frontal, no lateral ciliary tufts. Eye frontal, disc-shaped, red. Retrocerebral sac and subcerebral glands present. Oesophagus with inflation prior to entry of stomach. Gastric glands large, rounded. Pedal glands large, elongated, extending into trunk.

Incus small. Rami rhomboid. Fulcrum in ventral view rod-shaped, posterior end slightly expanded. Unci 5-toothed; principal tooth with prominence on ventral edge. Manubria long, needle-shaped and incurved posteriorly. Epipharyngeal elements bifurcate, large.

Length 190-213 μm, width 62-81 μm, toe 31-34 μm; manubrium 20 μm.

Distribution and ecology

Baltic and North Sea, Channel (France). Parasitic on gills of hermit crab (*Eupagurus bernhardus* (L.)); feeding on gill epithelium.

7. *Proales oculata* Tzschaschel, 1978

Figs. 102-105

Tzschaschel *in* Koste 1978: 272, pl. 89a figs. 4a-d; Tzschaschel 1979: 6-8, figs. 1a-d.

Type locality

Outside of Lister Hakens, Sylt, North Sea, Germany.

Description

Body vasiform in dorsal view; cuticle slightly stiffened. Body cavity with evenly shaped, small granulations; frequently also a single, large, globular accumulation of small granules. Head quadratic, distinctly offset by neck-fold. Trunk broadly elliptic, flattened ventrally, fairly arched dorsally. Tail distinct, blunt. Foot clearly offset, long, c. 1/4 total length, 3 pseudosegments with very small joint between pseudosegment 2 and 3. Toes relatively stout, close-set, laterally slightly arched, tapering to acute points, proximally with reservoir. Corona almost frontal, no lateral tufts of long cilia. Brain large, saccate. Y-shaped duct of rudimentary retrocerebral organ on brain; sac an indistinct accumulation of granules; subcerebral glands medium large. Frontal eye with 2 close-set pigment spots. No constriction between stomach and intestine. Gastric glands spherical. Pedal glands elongated, in trunk, ducts broad, always filled with secretion.

Rami triangular, anterior margins with 4 blunt teeth. Fulcrum short, rod-shaped in ventral view, posterior end knobbed; laterally wedge-shaped, slightly expanding posteriorly. Unci single-toothed, right 2, left a single accessory toothlet near tip. Manubria long, slender, posterior end incurved at a right angle, head slightly expanded with small, single lamella.

Length 140 μm, toe 17 μm; trophi 16 μm: ramus 5-6 μm, fulcrum 3 μm, uncus 8 μm, manubrium 13 μm.

Distribution and ecology

North Sea (Germany). Littoral psammon; April, May, August, sporadic, 18 °C. Literature: Tzschaschel (1980).

8. *Proales halophila* (Remane, 1929)

Figs. 106-110, pl. 6 figs. 1-6

? non *P. halophila* after Tzschaschel (1979)

Remane 1929b: 136, figs. 158a,b (*Proales globulifera* var. *halophila*); Bērziņš 1952: 6 (*Proales halophila*); Kutikova 1970: 486, figs. 69a,b; Koste 1978: 274, pl. 86 figs. 11a,b, pl. 89a figs. 1a-d; ? Tzschaschel 1979: 12-15, figs. 4a-e, figs. 5a-e.

Type locality
Not specified; North and Baltic Sea, Germany.

Description
Body vasiform in dorsal view; semiloricate. Body cavity with numerous light-refracting bodies, especially near gastric glands and nephridia. Head offset by neck-fold, ± quadratic, dorsally with delicate transverse fold. Trunk broad elliptic; flattened ventrally, dorsally fairly arched; distal end variable: acute, blunt, rounded with lateral ± deep indentations, and all possible transitions between. Foot moderately long, c. 2/5-1/3 total length; 3 pseudosegments, terminal longest, with small joint between segment 2 and 3, median pseudosegment with dorsal transverse fold. Toes long, slender; inner margins straight, outer margins curved, tapering to acute points. Corona slightly oblique, lateral ciliary tufts absent. Brain large; Y-shaped retrocerebral duct distinct. Two large subcerebral glands. Two frontal eyespots, sometimes absent. A shallow constriction between stomach and intestine. Gastric glands small, pyriform, stalked. Pedal glands elongate or spherical, extending into trunk, ducts very long; 2 reservoirs in anterior half of terminal foot pseudosegment.

Rami asymmetrical, each with triangular basal part and rectangular, dorsally recurved anterior part with thin, finger-shaped projection on its free end; alulae acute; ventrally each ramus with at least 3 crenated crests, right largest; large, crenated basal apophyses; fenestrae on dorsal and caudal margins. Fulcrum rod-shaped in ventral view, slightly expanding posteriorly; in lateral view ± tapering to oblique end. Uncus with 4-5 clubbed teeth, gradually decreasing in size, smallest with lamella; principal teeth with prominence on ventral edge. Manubria long, left somewhat longer, posterior end incurved at a right angle; head with short posterior and longer anterior cavity.

Length 135-180 μm, toe 18-22 μm; trophi 15-30 μm: ramus 5 μm, fulcrum 5 μm, uncus 9 μm, manubrium 12 μm.

Distribution and ecology
Baltic, North and Black Sea, Sea of Asow, ?Lake Baikal. In psammon and among algae; perennial, 1-22 °C. Literature: Czapik (1952).

Note
P. halophila (figs. 111-115) described by Tzschaschel (1979) from the German coast of the North Sea, differs by the absence of the small joint between pseudosegment 2 and 3, and the short terminal and long median pseudosegment; rami tips curved ventrally, manubria shorter.

Eggs ± rectangular, corners drawn out, underside concave, upper side convex (fig. 115).

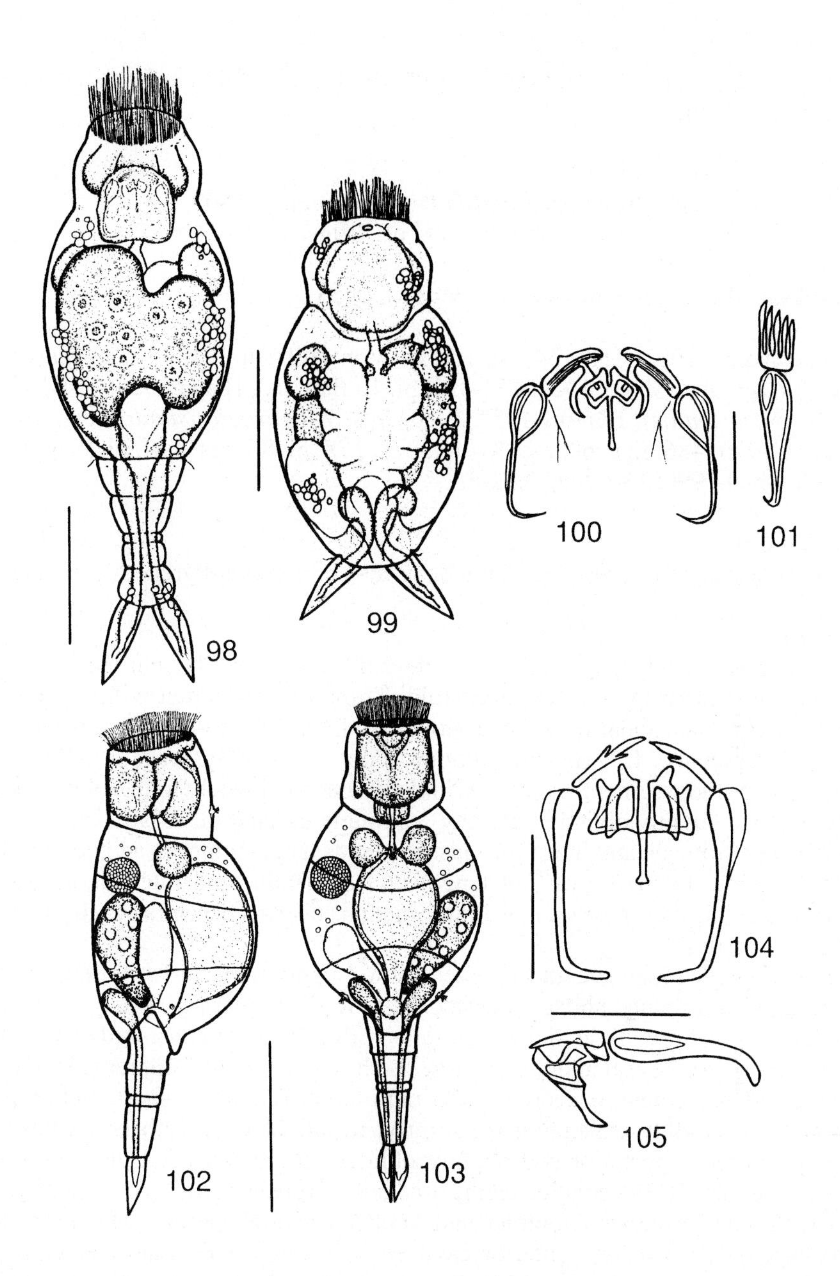

Figs. 98-101: *Proales paguri*. 98: ventral view, 99: dorsal view, 100: trophi, ventral view, 101: malleus. Figs. 102-105: *Proales oculata*. 102: lateral view, 103: dorsal view, 104: trophi, ventral view, 105: trophi, lateral view.
Scale bars: habitus 50 μm, trophi 10 μm.
(98-101: after Thane-Fenchel, 1966; 102-105: after Tzschaschel, 1979)

Length 135 μm, toe 22μm; trophi 21 μm: ramus 5 μm, fulcrum 5 μm, uncus 9 μm, manubrium 12 μm.

9. *Proales similis* De Beauchamp, 1907

Figs. 116-128, pl. 7 figs. 1-6

Infrasubspecific taxon: var. *exoculis* Bērziņš, 1953

De Beauchamp 1907: 153-154, figs. 2a,b; Von Hofsten 1912: 173 (*Pleurotrocha similis*); Harring & Myers 1924: 434-435, pl. 16 figs. 1-5; Hauer 1925: 154-155, figs. 1a,b (*Proales similis*); Bērziņš 1953: 9-10, figs. 7-9 (*Proales similis* var. *exoculis*); Kutikova 1970: 486, figs. 695a-e; Koste 1978: 278, pl. 86 figs. 4a-f, pl. 91 figs. 5a-c; Koste & Shiel 1990: 136, figs. 3.2a-d.

Type locality
Salt marshes near fort of Socoa, Saint-Jean-de-Luz (Basses-Pyrenées), France.

Description
Body fusiform, elongate, slender; cuticle soft, very flexible, outline ± variable; hyaline. Head offset by discrete neck-fold. Trunk dorsally often with symmetrical, longitudinal folds; caudal region occasionally with 2-3 latero-dorsal transverse folds or pseudosegments. Tail distinct, minute. Foot moderately long, c. 1/5-1/6 total length, one pseudosegment, usually wrinkled. Toes moderately long, robust, conical, acute. Corona slightly oblique, laterally 2 strongly ciliated tufts. No distinct separation between stomach and intestine. Gastric glands large, compressed laterally. Pedal glands elongate to pyriform, foot length or less. Brain moderately large, saccate. Eyespot red, large, at posterior of brain, medial, displaced ventrally. Retrocerebral sac small, ductless.

Trophi between malleate and virgate. Rami roughly trapezoid with long, dorsally curved tips, and acute alulae; anterior half of inner margins finely denticulate, dorsally a second row of denticles; ventrally with 3 crests of appressed teeth anterior to basal apophyses; basal apophyses large, each with comb of 7 appressed teeth; dorsal fenestrae prominent, ± central; bulla rami large; fulcrum ± short, rod-shaped in ventral view, slightly expanding posteriorly; in lateral view with broad base, narrower posteriorly, posterior end slightly oblique and expanded. Unci with 4 large teeth and group of 3-4 smaller teeth, gradually decreasing in size; principal teeth each with small accessory toothlet and knob on ventral margin. Manubria broad, club-shaped, posterior and anterior cavities long and broad, almost extending till curved end of manubria. Epipharynx 2 long, slender, slightly curved rods.

Length 110-180 μm, toe 7-20 μm; trophi 18-24 μm: ramus 15 μm, fulcrum 5-7 μm, uncus 7-12 μm, manubrium 15-19 μm.

Distribution and ecology
Europe, N. America, Australia, ?Indonesia (Borneo). Halophile, in inland saline, marine and brackish waters; salinity up to 98‰ , conductivity up to 163.0 $mScm^{-1}$, pH 7.4-8.3. Literature: Brain & Koste (1993).

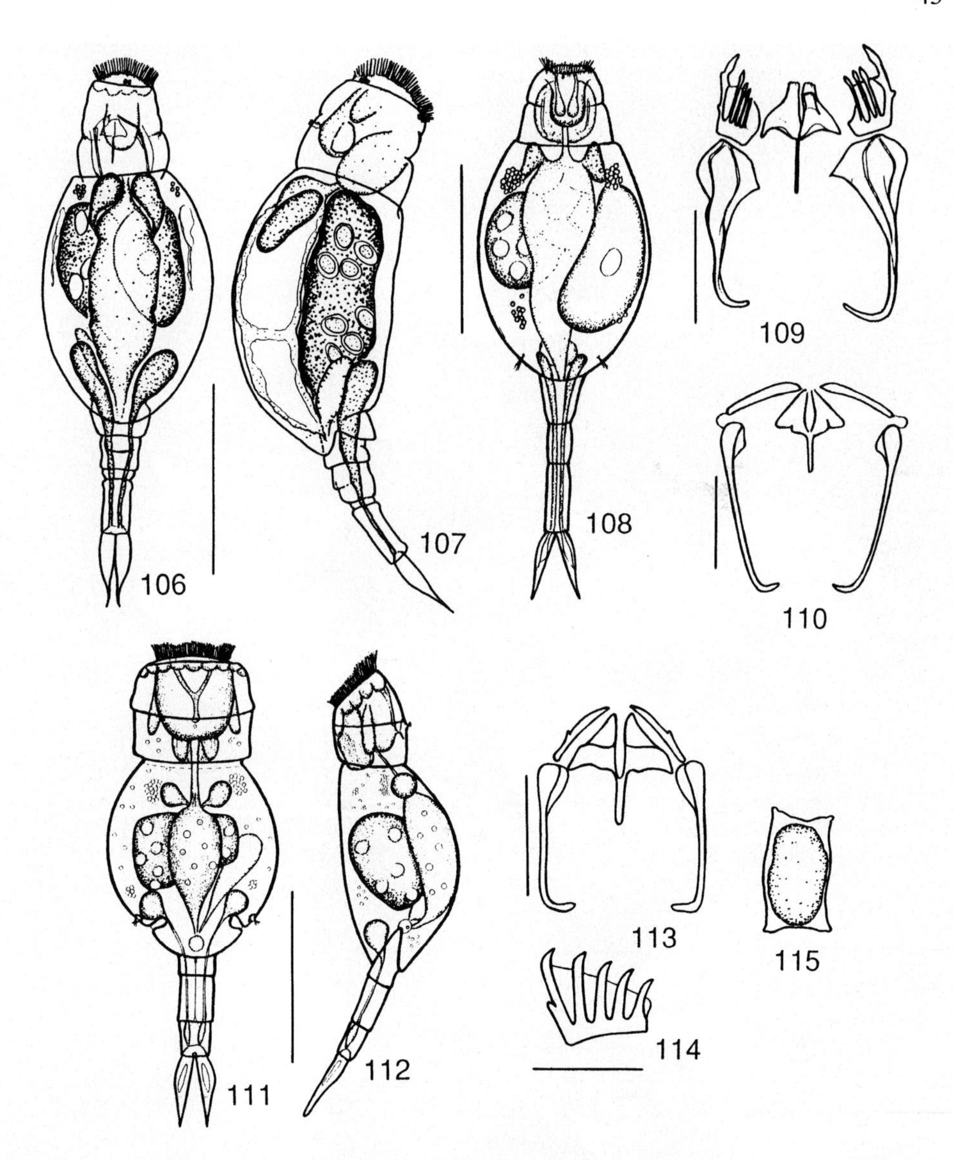

Figs. 106-110: *Proales halophila*. 106: dorsal view, 107: lateral view, 108: dorsal view, 109: trophi, ventral view, 110: trophi.
Figs. 111-115: *Proales halophila* sensu Tzschaschel. 111: dorsal view, 112: lateral view, 113: trophi, ventral view, 114: right uncus, 115: egg.
Scale bars: habitus 50 µm, trophi 10 µm (114: 5 µm).
(106: Oesterdam, Eastern Scheldt, The Netherlands; 107,109 Trygghamna, Spitsbergen, Svalbard; 108,110: after Remane, 1929b; 111-115: after Tzschaschel, 1979)

Plate 6. *Proales halophila*, trophi (S.E.M. photographs). 1: lateral view, 2: ventro-apical view, detail incus, 3: idem, lateral view, 4: ventro-caudal view, 5: caudal view, detail incus, 6: idem, ventral view. Scale bars: 10μm.
(1-6: Channel, Audresselles, France)

10. *Proales theodora* (Gosse, 1887)
Figs. 129-134, pl. 8 figs. 1-3

Synonym: ? *P. longipes* Remane, 1929

Infrasubspecific taxon: var. *calcarata* Wulfert, 1938

Gosse 1887b: 862, pl. 14 fig. 2 (*Notommata theodora*); ? Remane 1929b: 151 (*Proales longipes*); Hauer 1938: 214-217, figs. 1a-g (*Proales theodora*); Wulfert 1938: 385-388, figs. 160a-f, 161 (var. *calcarata*); Donner 1955: 103-104, figs. 32a-c; Kutikova 1970: 488, figs. 700a-f; Koste 1978: 270-271, pl. 89 figs 2a-g; Braioni & Gelmini 1983: 87, 89, figs. 42a-e, 43a-c.

Type locality
Not specified, England.

Description
Body fusiform, elongate, slender; hyaline. Head offset by distinct neck-fold, completely retractable. Tail small, prominent. Foot very long, ≥ 1/3 total length, very motile, 3 pseudosegments: basal short, ± conical, median long, ± cylindrical, terminal very short, usually partially retracted in median. Toes ± lanceolate, abruptly ending in tubular points, bases partially fused, often slightly recurved dorsally. Corona slightly oblique, 2 lateral ciliary tufts. Brain large, saccate. Two frontal eyespots, x-like set together, occasionally fused. Stomach and intestine separated by constriction. Gastric glands oval. Pedal glands very long, extending into trunk.

Trophi malleate, symmetrical; basal part of rami with rounded alulae and large, plate-shaped basal apophyses with comb of ± 12 teeth; anterior part thin, Y-shaped. Fulcrum short, posterior end strongly expanded in ventral view; in lateral view ± wedge-shaped, posterior end slightly expanded. Unci symmetrical, 3-4 rounded teeth, gradually decreasing in size; principal teeth each with prominence on ventral margin; under side with skleropili on anterior margin of uncinal plates. Manubria club-shaped, wide, posterior half curved ventrally, posterior and anterior cavities in upper half. A thin ± triangular plate between malleus and incus. Occasionally 2 thin, kinked, pleural rods.

Subitaneous egg elongate oval, attached to substrate by short tag.

Length 265-500 μm, toe 27-39 μm; trophi 22-29 μm: ramus 14-18 μm, fulcrum 6 μm, uncus 10-12 μm, manubrium 23-26 μm; subitaneous egg 88-116x50-62 μm.

Animals differing by their more slender toes with demarcated tips, and triangular frontal eye, have been described as *P. longipes* var. *calcarata* by Wulfert (1938).

Distribution and ecology
Europe, America, New Zealand. In mountain springs and littoral of alkaline stagnant and running fresh waters, also among algae in littoral of brackish and marine waters. Food: diatoms, detritus, bacteria, dinoflagellates. Literature: Pejler (1962), Donner (1964), Thane-Fenchel (1968).

11. *Proales reinhardti* (Ehrenberg, 1834)
Figs. 135-141, pl. 9 figs. 1-6

Ehrenberg 1834: 208 (*Furcularia reinhardti*); Harring & Myers 1924: 431-434, pl. 16 figs. 6-10 (*Proales reinhardti*); Remane 1929-33: 540, figs. 321a-f; Hollowday 1949b: 248, 250, figs. 2d,e; Koste 1970: 270, pl. 86 figs. 9a,b,d-g, pl. 89 figs. 3,7a-d; Kutikova 1970: 488, figs. 697a-e; Chengalath 1985: 2214-2215, fig. 5.

Type locality
Baltic Sea, Germany.

Description
Body fusiform, elongate, slender; cuticle very flexible, outline variable; hyaline. Head offset by distinct neck-fold. Tail prominent, rounded. Foot very long, ≥ 1/4 total length, 2 pseudosegments, basal pseudosegment short, stout, c. 1/3 length slender terminal pseudosegment; foot telescopically retractable within body. Toes long, slender, lanceolate. Corona slightly oblique, 2 lateral ciliary tufts. Brain moderately large, saccate. Two frontal eyespots, set close together, occasionally fused, often disappearing in adults. Retrocerebral organ absent. Stomach and intestine separated by shallow constriction; proventriculus present. Gastric glands rounded, compressed laterally. Pedal glands very long, elongated, extending into trunk.

Rami asymmetrical, broadly triangular, short, alulae short, ± acute; inner margins right with 1-(2) pair(s) of close set teeth; left basal apophyse very large, right short, small; dorsal fenestrae postero-lateral, right larger; a single ventral fenestra, left. Fulcrum very short, in venral view rod-shaped with slightly expanded posterior end, in lateral view slightly curved. Unci with 4-5 clubbed teeth, ± gradually decreasing in size, principal teeth each with accessory toothlet ventrally; inner side with preunci and plate with skleropili. Manubria long, slender, clubbed, head with 2 short lateral cavities; opening of posterior cavity on inner side, opening of median and anterior cavity externally; posterior end incurved.

Length 200-380 μm, toe 20-28 μm; trophi 30-45 μm: ramus 9-12 μm, fulcrum 6 μm, uncus 12-14 μm, manubrium 24-25 μm.

Male (fig. 137) with functional (?) mastax; foot short, toes lanceolate with slightly drawn-out tips, decurved ventrally. Length 150-155 μm.

Distribution and ecology
Wide-spread: Baltic, North and Black Sea, Sea of Asow, N. American Atlantic, Arctic (Barents and Laptev Sea), Antarctic. In vegetation and sedimentation zone, rock pools, marine and brackish tide puddles, in brine channels of arctic sea ice; -1.5-20°C, salinity 6-40‰. Food: dinoflagellates, small diatoms, bacteria. Literature: Björklund (1972), Budde (1924), Hauer (1938), Jansson (1967), Thane-Fenchel (1968).

Note
The species is often confused with *P. theodora*. All records of *P. reinhardti* from freshwater are doubtful.

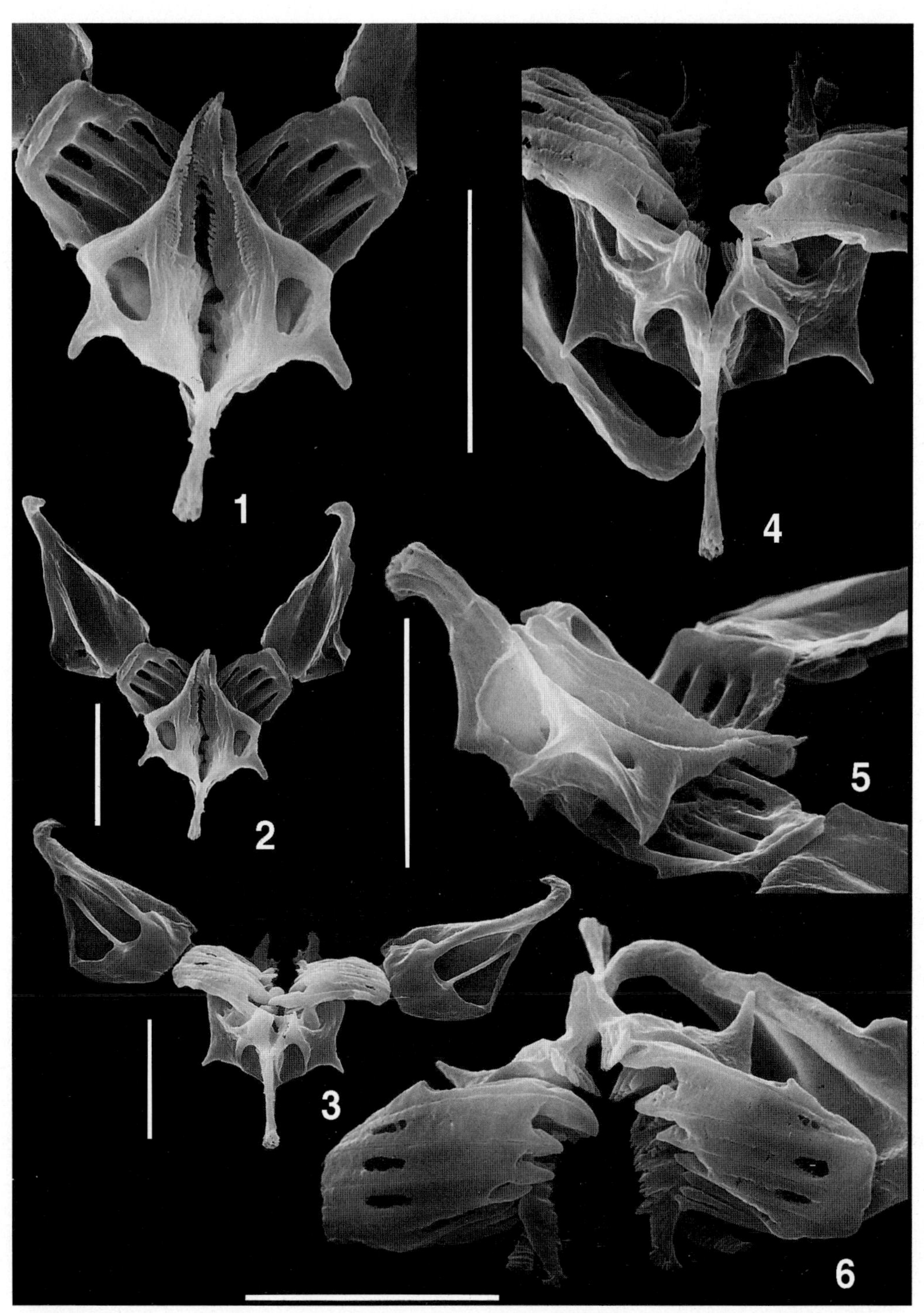

Plate 7. *Proales similis*, trophi (S.E.M. photographs). 1: dorso-caudal view, detail incus, 2: dorso-caudal view, 3: ventro-apical view, 4: idem, detail incus, 5: lateral view, detail incus, 6: apical view, detail incus and unci. Scale bars: 10μm. (1-6: Mediterranean, Port Vendres, France)

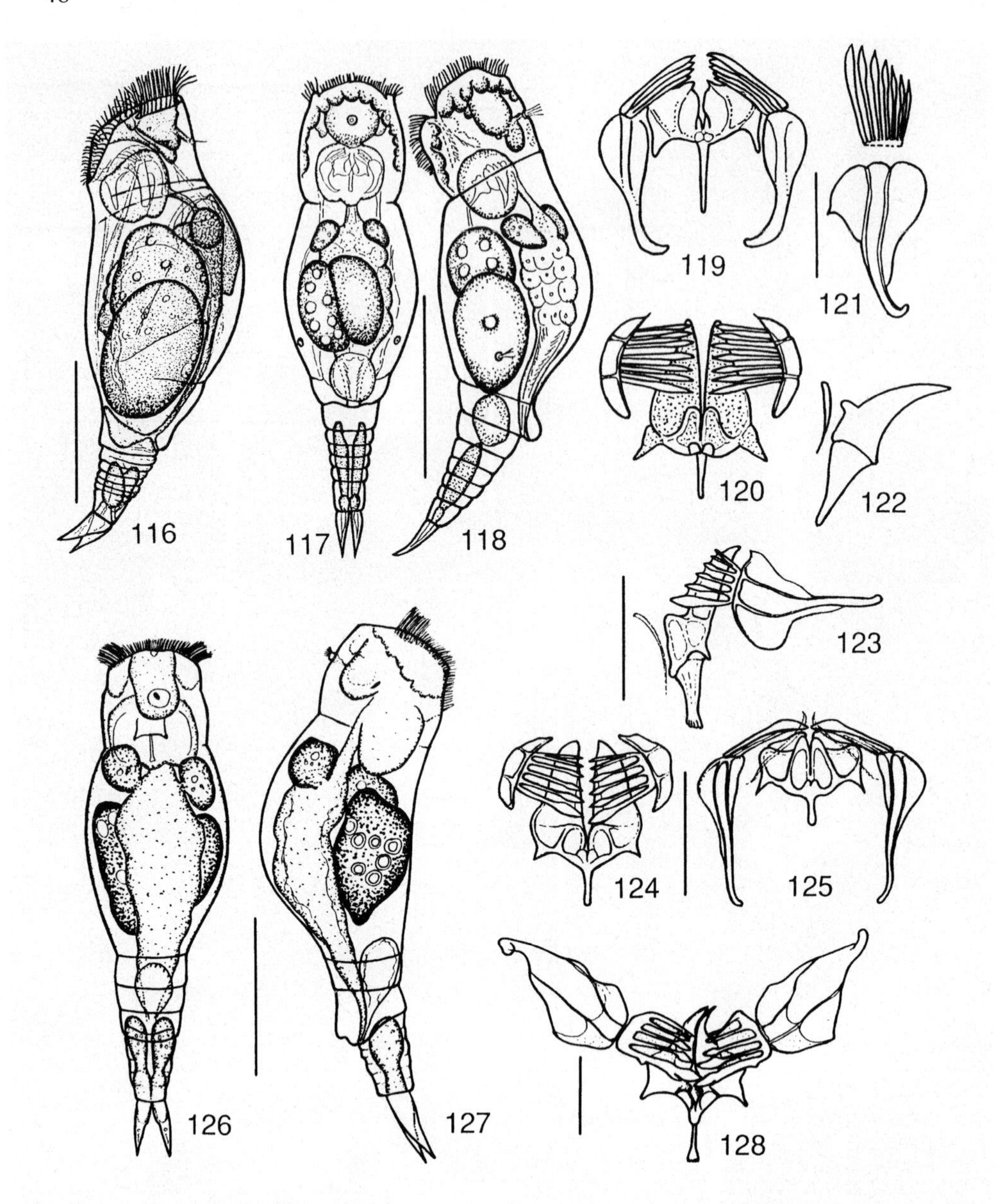

Figs. 116-128: *Proales similis*. 116: lateral view, 117: ventral view, 118: lateral view, 119: trophi, ventral view, 120: trophi, apical view, 121: malleus, 122: incus with epipharynx, lateral view, 123: trophi, lateral view, 124: trophi, apical view, 125: trophi, ventral view, 126: dorsal view, 127: lateral view, 128: trophi, ventro-apical view.
Scale bars: habitus 50 μm, trophi 10 μm.
(116: after De Beauchamp, 1907; 117-122: after Koste & Shiel, 1990; 123-125: after Harring & Myers, 1924; 126-128: Mediterranean, Port Vendres, France)

Plate 8. *Proales theodora*, trophi (S.E.M. photographs). 1: ventro-apical view, 2: idem, lateral view, 3: dorso-caudal view (left malleus outer side, right malleus inner side).
Scale bar: 10 μm.
(1,2: North Sea, Nieuwpoort, Belgium; 3: Torgny, Luxemburg, Belgium)

12. *Proales globulifera* (Hauer, 1921)
Figs. 142-148

Synonym: ? *P. quadrangularis* Glascott, 1893

Hauer 1921: 184-185, text fig. (*Furcularia globulifera*); ? Glascott 1893: 43, pl. 3 fig. 3 (*Notops quadrangularis*); Remane 1929b: 136 (*Proales globulifera*); Schulte 1959: 181-182, figs. 5a-c; Donner 1964: 296, figs. 35a-c; Kutikova 1970: 487-488, figs. 696a,b; Koste 1976: 210, pl. 19 fig. 2, pl. 21 figs. 6a,b; Koste 1978: 272-273, pl. 86 figs. 10a,b, pl. 89 figs. 4,5a-c; Jersabek & Schabetsberger 1992a: 98, figs. 39a-c.

Type locality
Quarry near Donaueschingen, Germany.

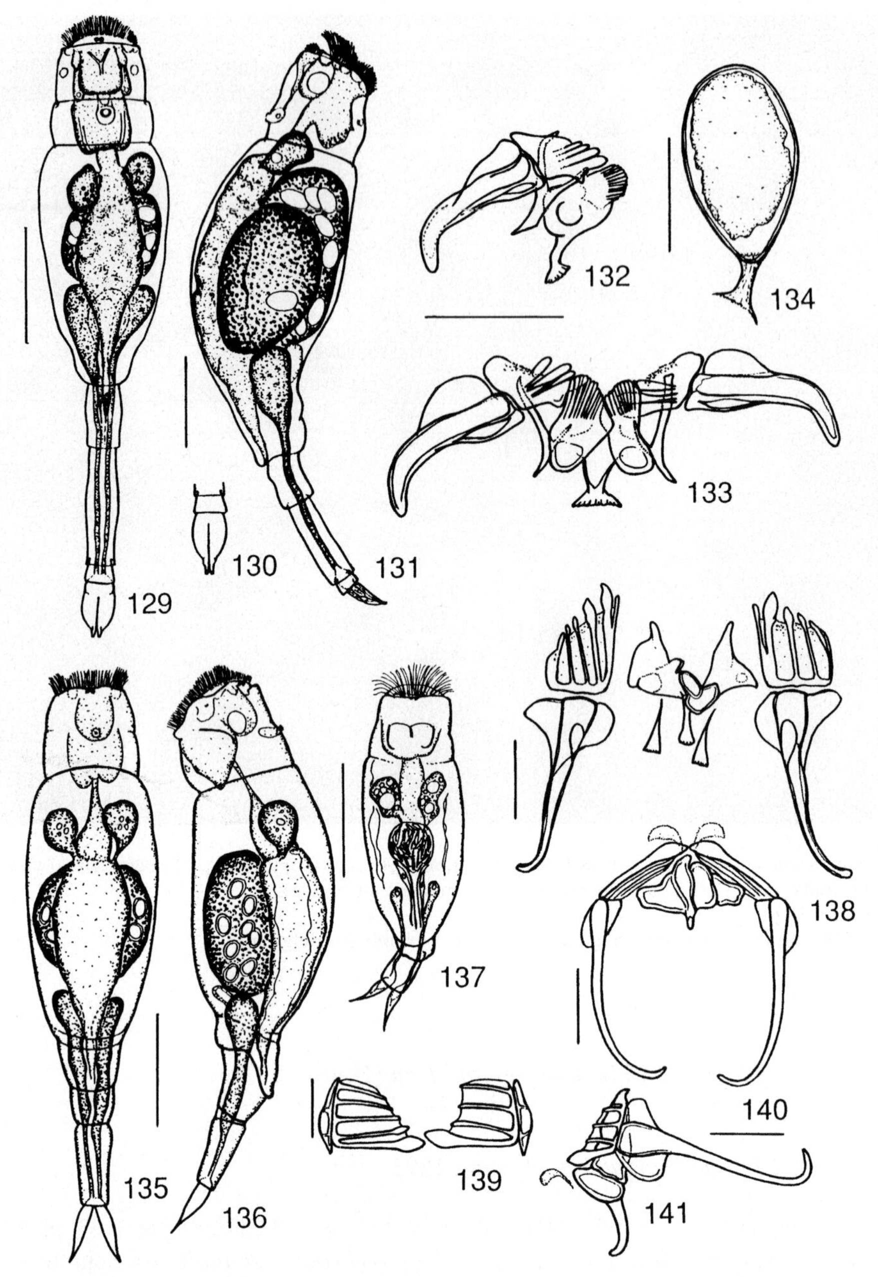

Figs. 129-134: *Proales theodora*. 129: dorsal view, 130: toes, dorsal view, 131: lateral view, 132: trophi, lateral view, 133: trophi, ventral view, 134: subitaneous egg.
Figs. 135-141: *Proales reinhardti*. 135: dorsal view, 136: lateral view, 137: male, dorsal view, 138: trophi, ventral view, 139: unci, 140: trophi, ventral view, 141: trophi, lateral view.
Scale bars: habitus and egg (134) 50 μm, trophi 10 μm.
(129,130,132: Tempelfjorden, Spitsbergen, Svalbard; 131,134: Trygghamna, Spitsbergen, Svalbard; 133: Torgny, Belgium; 135: Oesterdam, Eastern Scheldt, -9 m, The Netherlands; 136: Colijnsplaat, littoral Eastern Scheldt, The Netherlands; 137,138: arctic sea ice, Laptev Sea, leg. C. Friedrich; 139-141: after Harring & Myers, 1924)

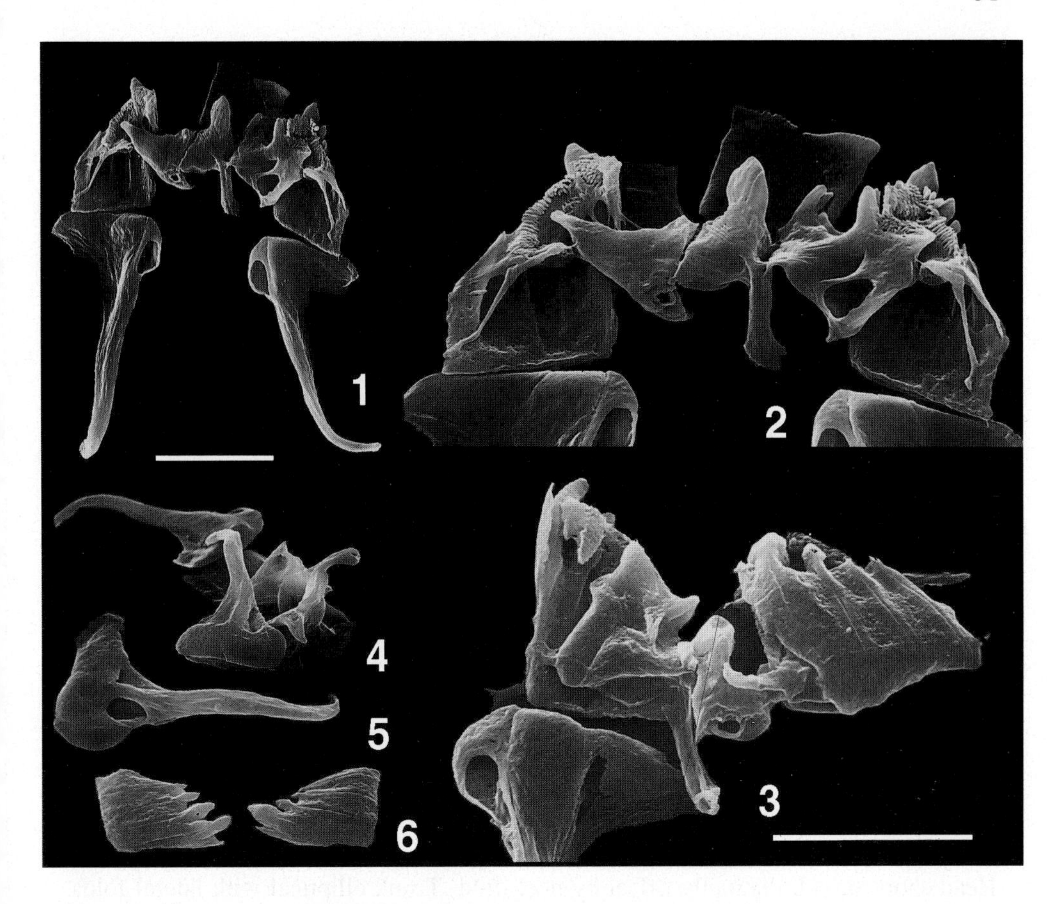

Plate 9. *Proales reinhardti*, trophi (S.E.M. photographs). 1: dorsal view,2: idem, detail, 3: ventral view, detail, 4: lateral view, 5: right manubrium, 6: unci.
Scale bars: 10μm.
(1,2: Eastern Scheldt, Colijnsplaat, The Netherlands; 3-6: arctic sea ice, Laptev Sea, leg. C. Friedrich)

Description

Body vasiform in dorsal view; semiloricate; transparent. Body cavity with 4 symmetrically placed, spherical accumulations of light-refracting bodies. Head offset by neck-fold, quadratic. Trunk elliptical, dorsally arched, ventrally flattened; longitudinal folds laterally. Head and foot retractable in trunk. Foot long, c. 1/4-1/5 total length, slender, 2-3 pseudosegments, terminal pseudosegment longest. Toes long, slender, lanceolate. Corona simple, almost frontal, ciliary tuft above mouth. Two frontal, red eyespots, sometimes fused; crystalline body present or not. No constriction between stomach and intestine. Gastric glands very large, rounded. Pedal glands elongate, extending into trunk.

Mastax very large. Incus small. Rami triangular, lateral ± blunt alulae. Fulcrum very short, rod-shaped in ventral view. Unci with one principal and 4 subsidiary teeth; ventral edge of principal tooth with blunt prominence. Manubria long, slightly asymmetrical, incurved posteriorly, short posterior and longer anterior lamella.

Length 130-200 μm, toe 21-25 μm; trophi 21-23 μm: fulcrum 5 μm, left manubrium 17 μm, right manubrium 15-17 μm, uncus 8-9 μm; spherical accumulation of light-refracting bodies up to 1.6 μm.

Distribution and ecology
Europe. In limno- and potamopsammon, in algal films; mainly in the cold seasons. Food: small diatoms.

13. *Proales minima* (Montet, 1915)
Figs. 149-153, pl. 10 figs. 1,2

Synonym: ? *P. psammophila* Neiswestnowa-Shadina, 1935

Montet 1915: 323-324, pl. 13 figs. 33a-d (*Pleurotrocha minima*); Weber & Montet 1918: 103 (*Proales minima)*; Harring & Myers 1924: 435, pl. 20 figs. 1-4; Wiszniewski 1934b: 342, pl. 58 fig. 1; ? Neiswestnowa-Shadina 1935: 560-561, figs. 5,6 (*P. psammophila*); Wiszniewski 1935: 229-230; Donner 1952a: 15-16, figs. 7a-d; Kutikova 1970: 488, figs. 699a-d; Koste 1976: 210, pl. 21 figs. 5a,b; Koste 1978: 273, pl. 86 figs. 12a-d, pl. 89a figs. 5a-d; Jersabek & Schabetsberger 1992a: 98, figs. 38a-f.

Type locality
In moss from pond Issalets, Bassin de Léman, Switzerland.

Description
Body vasiform in dorsal view; cuticle slightly stiffened, outline constant; hyaline. Head short, broad, distinctly offset by neck-fold. Trunk elliptical with lateral folds. Tail short. Foot long, c. 1/4-1/5 total length, slender, 3 pseudosegments. Toes fairly long, slender, nearly cylindrical half their length, tapering to acute points, slightly decurved ventrally. Corona slightly oblique, lateral ciliary tufts like auricles. No constriction between stomach and intestine. Gastric glands small, rounded. Pedal glands minute. Brain fairly large, saccate. Retrocerebral sac rudimentary. Eyespot very small, frontal, sometimes absent.

Rami broadly triangular, apically with strongly narrowing prominences; lateral alulae acute; inner margins ventrally with large crenate crests, anterior to the very large, crenated basal apophyses. Fulcrum rod-shaped in ventral view, laterally slightly narrowing to posterior end. Unci with 4-5 teeth, smallest teeth each with lamella, principal teeth each with short prominence on ventral edge. Manubria club-shaped, posterior end slightly incurved, lateral lamellae small, anterior one with recurved corner; median cavity with small, drop-shaped aperture at outer side. Epipharynx 2 fairly large, triangular plates.

Length 70-120 μm, toe 12-30 μm; trophi 9-15 μm: ramus 5 μm, fulcrum 2-3 μm, uncus 6 μm, manubrium 9-10 μm.

Distribution and ecology
Europe, N. America. In submerged mosses, limno- and potamopsammon.

14. *Proales doliaris* (Rousselet, 1895)

Figs. 154-160, pl. 11 figs. 1-4

Rousselet 1895: 120, pl. 7 fig. 4 (*Microcodides doliaris*); Harring & Myers 1924: 437-438, pl. 19 figs. 3-7 (*Proales doliaris*); Wulfert 1940: 583, figs. 29,29b; Koste 1965: 72, fig. 16; Kutikova 1970: 490, figs. 703a-d; Martin 1977: 239, fig. 3(male); Koste 1978: 269-270, pl. 87 figs. 1a-e; Koste & Shiel 1990: 135, figs. 2.2a-d; Jersabek & Schabetsberger 1992a: 99, figs. 40a-f.

Type locality
United Kingdom.

Description
Body short, stout, vasiform; cuticle soft, very flexible, outline constant; hyaline. Head broad, short, slightly tilted ventrally, neck-fold absent. Trunk ovoid or spherical, dorsally 5-6 indistinct transverse folds. A short, sleeve-like tail surrounding base of foot. Foot ≤ 1/10 total length, 2-3 pseudosegments. Toe single, fusiform, acute. Corona strongly oblique, a circumapical band of ± short cilia, 2 lateral ciliary tufts, apical field naked with 2 papillae. Brain very large, saccate. Retrocerebral organ absent. Eyespot red, ventrally on brain. A constriction between stomach and intestine. Gastric glands spherical. Two pyriform pedal glands.

Rami with triangular basal part and ± rectangular anterior part; inner margins of rectangular part with simple, long, slender toothlets; inner margins of basal part with 3-4 conical teeth, surrounded by appressed spines; right ramus with lamellar projection near base, with 5-6 marginal teeth; basal apophyses long, conical, each with large, rounded opening of bulla rami at its base. Fulcrum ± rod-shaped in ventral view, slightly expanding posteriorly; in lateral view broad, ± parallel-sided, slightly narrowing posteriorly. Right uncus with 7, left with 6 acute, slightly clubbed teeth, gradually decreasing in size; inner side with long skleropili near anterior margin. Manubria moderately long, club-shaped each with short posterior cavity and longer, lamellar anterior cavity. Epipharynx 2 fairly large, triangular plates.

Length 170-300 μm, toe 14-25 μm; trophi 25-32 μm: ramus 15-21 μm, fulcrum 10-12 μm, uncus 13-17 μm, manubrium 20-22 μm.

Male (fig. 156) resembling female, with ringed foot and single toe. Length 116 μm, toe 17 μm.

Distribution and ecology
Wide-spread (Eurasia, N. America, Australia, New Zealand). In moor and soft waters; warm stenotherm, pH 4.8-6.5.

15. *Proales cryptopus* Wulfert, 1935

Figs. 161-164

Wulfert 1935: 601-602, figs. 16, 16a-c; Koste 1978: 279, pl. 87 figs. 3a-c; Kutikova 1970: 499, figs. 721a,b.

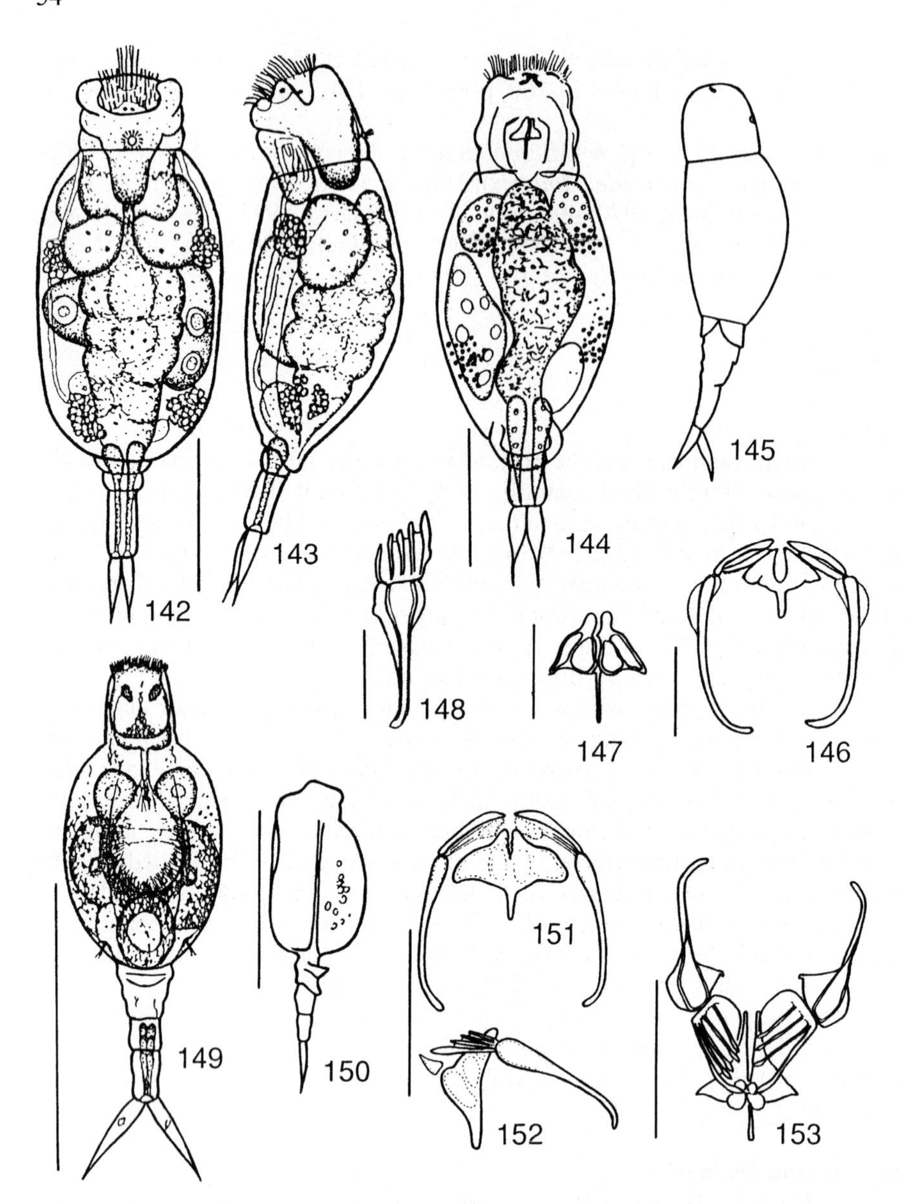

Figs. 142-148: *Proales globulifera*. 142: dorsal view, 143: lateral view, 144: dorsal view, 145: lateral view, 146: trophi, ventral view, 147: incus, dorsal view, 148: malleus, right.
Figs. 149-153: *Proales minima*. 149: dorsal view, 150: lateral view, 151: trophi, ventral view, 152: trophi, lateral view, 153: trophi, ventro-apical view.
Scale bars: habitus 50 μm, trophi 10 μm.
(142,143: after Hauer, 1921; 145,148: after Donner, 1964; 144,147: after Schulte, 1959; 146: after Koste, 1976; 149,150: after Donner, 1952a; 151,152: after Harring & Myers, 1924, 153: Canigou, Pyrenées Orientales, France)

Plate 10. *Proales minima*, trophi (S.E.M. photographs). 1: ventro-apical view, 2: lateral view.
Scale bar: 10μm.
(1,2: Canigou, Pyrenées Orientales, France)

Type locality
Merseburg, Halle, Germany.

Description
Body cylindrical, head and posterior rounded; very transparent, almost colourless (stomach light-yellowish). Head offset by neck-fold; rostrum small. Trunk with pseudosegments of different length. Foot short. Toes very short, 1/34 total length, conical, covered by tail. Corona?, cilia short, apparently without ciliary tufts. Stomach separated from intestine by deep constriction. Gastric glands very large, elongated with acute ends, ducts distinct. Pedal glands ovoid, with a single small reservoir. Brain large, saccate. Retrocerebral sac present. Eyespot red, on brain placed to the right.

Mastax quadratic. Rami complex: two broad plates, posterior margins obliquely tapering towards fulcrum, anterior margins broadly rounded with 2 teeth; basal apophyses large. Fulcrum short, gradually expanding towards posterior end in ventral view. Unci with 4 teeth, principal bifurcate. Manubria with single lamella ? Two S-shaped pleural rods.

Length 375 μm, toe 11 μm; trophi 25 μm: fulcrum 11 μm, rami width 20 μm, uncus 14 μm.

Distribution and ecology

Europe (Germany), S. America (Surinam). Periphytic in ponds; January, pH 7.5-8.

16. *Proales macrura* Myers, 1933

Figs. 165-169

Myers 1933a: 24-25, figs. 14a-e.

Type locality

Duck Brook, Mount Desert Island, Maine; Cordroy Creek, Atlantic County, New Jersey, U.S.A.

Description

Body cylindrical, elongate; cuticle flexible, outline variable. Head offset by neck-fold, fairly long, c. 1/3 total length; rostral process prominent. Tail prominent, over-hanging. Foot placed ventrally, tubular, short c. 1/7 total length; 3 pseudosegments. Toes short, bases slightly swollen, tips fine, dorsally recurved. Corona almost ventral, 2 prominent lateral ciliary tufts, dorsal arc of circumapical band absent. Brain fairly large. Retrocerebral sac clear, ductless, posteriorly on brain. Eyespot small, at posterior of brain. No constriction between stomach and intestine. Gastric glands large, kidney-shaped, somewhat placed ventrally. Pedal glands very small, confined to terminal foot pseudosegment.

Trophi modified malleate. Rami triangular, inner margins without teeth; basal apophyses prominent. Fulcrum long, slender, rod-shaped. Left uncus with 3, right with 4 clubbed teeth. Manubria long, slightly bent; anterior lamella narrow, manubrium length, posterior one absent.

Length 149 μm, toe 9 μm; trophi 20 μm.

Distribution and ecology

N. America (Maine, New Jersey), Europe (Sweden). In marginal *Sphagnum* of acid brooks and ponds.

17. *Proales alba* Wulfert, 1939

Figs. 170-176

Wulfert 1939: 595-596, figs. 11a-h; Kutikova 1970: 499, figs. 719a-e; Koste 1978: 277-278, pl. 91 figs. 4a-h.

Type locality

Rockendorf, Beesen and Etzdorf near Halle, Merseburg, Germany.

Description

Body ± uniformly tapering towards toes, older specimens slightly expanded near middle of trunk; strongly hyaline, colourless, stomach and vitellarium yellowish-grey. Head offset by neck-fold. Trunk with one short and 3 large pseudosegments; numerous delicate, longitudinal folds. Tail small, distinct. Foot moderately long, c. 2/7-1/4 total length, 4 pseudosegments. Toes soft, flexible, acute, bases inflated, partially fused, tips slightly curved dorsally. Corona oblique, 2 lateral ciliary tufts. No constriction between stomach and intestine. Gastric glands relatively small, oval. Pedal glands foot length, elongate pyriform, with reservoirs. Brain quadratic, short. Retrocerebral sac hemispherical, with granules, hyaline. Eye on brain, red, large, median.

Rami anteriorly with 2 frontal, widely set apart projections; inner margins smooth; alulae blunt; basal apophyses rounded. Fulcrum straight, slender, gradually expanding towards posterior end in ventral view; laterally ± knife-shaped. Unci 2-toothed with rounded plate. Manubria ± straight, posterior and anterior lamella 1/2 manubrium length.

Length 140-190 μm, toe 10 μm; trophi 13 μm: ramus 7 μm, fulcrum 7 μm, manubrium 12 μm.

Distribution and ecology

Europe (Germany, Czechia), N. America (Canada). In ditches, ponds, etc.

18. *Proales bemata* Myers, 1933

Figs. 177-179

Myers 1933a: 21-22, figs. 12a-c.

Type locality

Mount Desert Island, Maine; Atlantic County, New Jersey, U.S.A.

Description

Body tapering towards toes; cuticle very soft, flexible, outline variable. Head large, offset by well-marked neck-fold. Trunk gradually tapering towards foot. Tail small, distinct. Foot long, c. 1/4 total length, 4 pseudosegments, ± equally long. Toes short, tips acute, bases slightly swollen; a septum divides tips from bases. Corona oblique, 2 lateral areas with long cilia. Brain large. Retrocerebral sac small, confluent, black, attached to dorsal side of brain. Eyespot on brain, displaced to right. Stomach and intestine small, not separated by constriction. Gastric glands very large, oval, vacuolated. Pedal glands stout, long, foot length.

Trophi modified malleate. Rami roughly triangular, inner margins without teeth, only a slight swelling on right ramus near mid-length; basal apophyses large. Fulcrum long, rod-shaped in ventral view, posterior end with fan-shaped expansion; laterally tapering from a broad base to an obtuse fan-like expansion. Unci with 5 clubbed teeth, gradually decreasing in size.

Manubria ± long, stout, slightly incurved, posterior end knobbed, posterior and anterior lamella 1/2 and 2/3 length of manubrium respectively. Epipharynx 2 elongate rods, bent at their extremities.

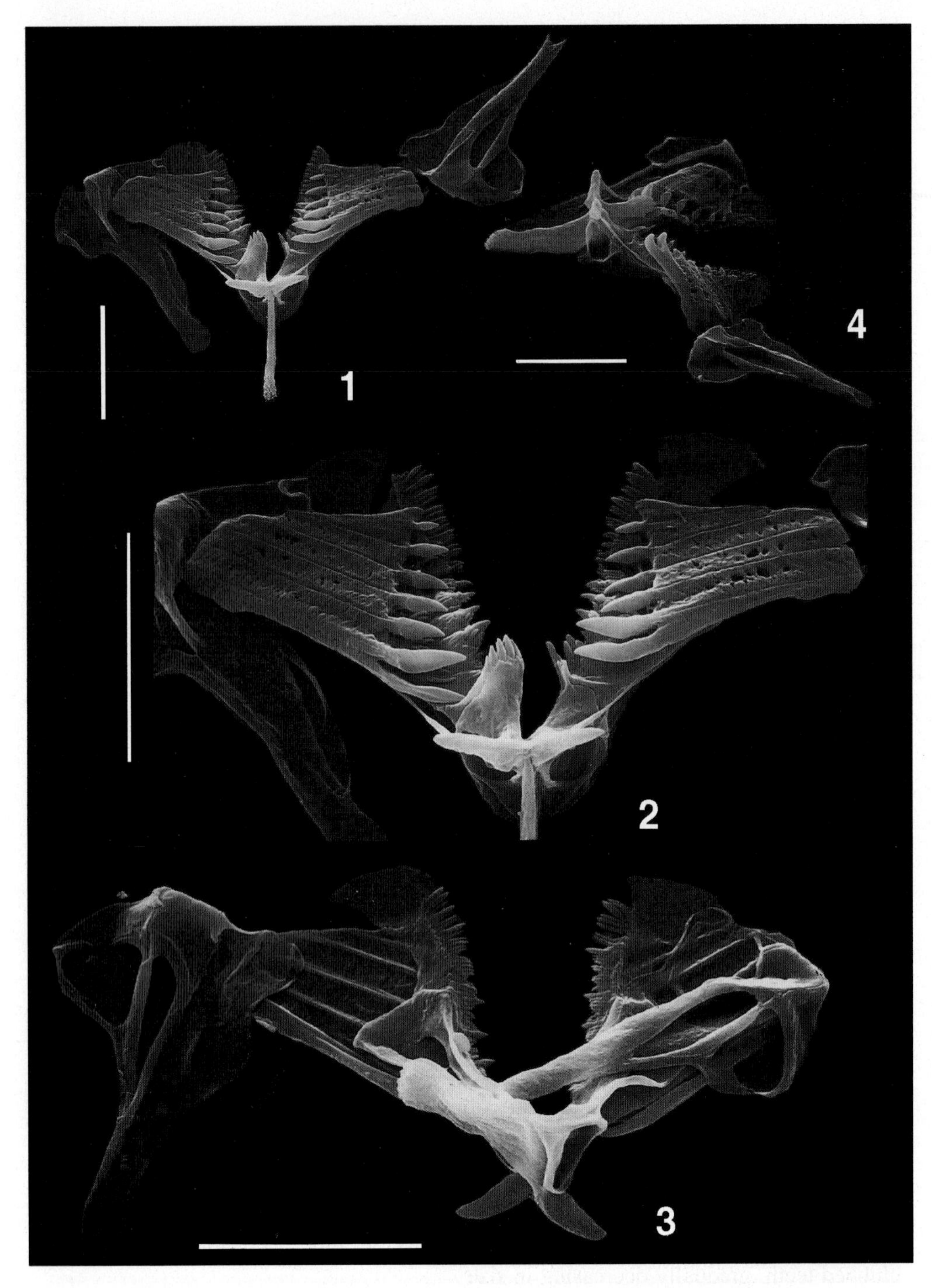

Plate 11. *Proales doliaris*, trophi (S.E.M. photographs). 1: ventro-apical view, 2: detail rami and unci, 3: caudal view, 4: lateral view.
Scale bars: 10μm.
(1-4: Hohen Tauern, Austria, leg. & det. C. Jersabek)

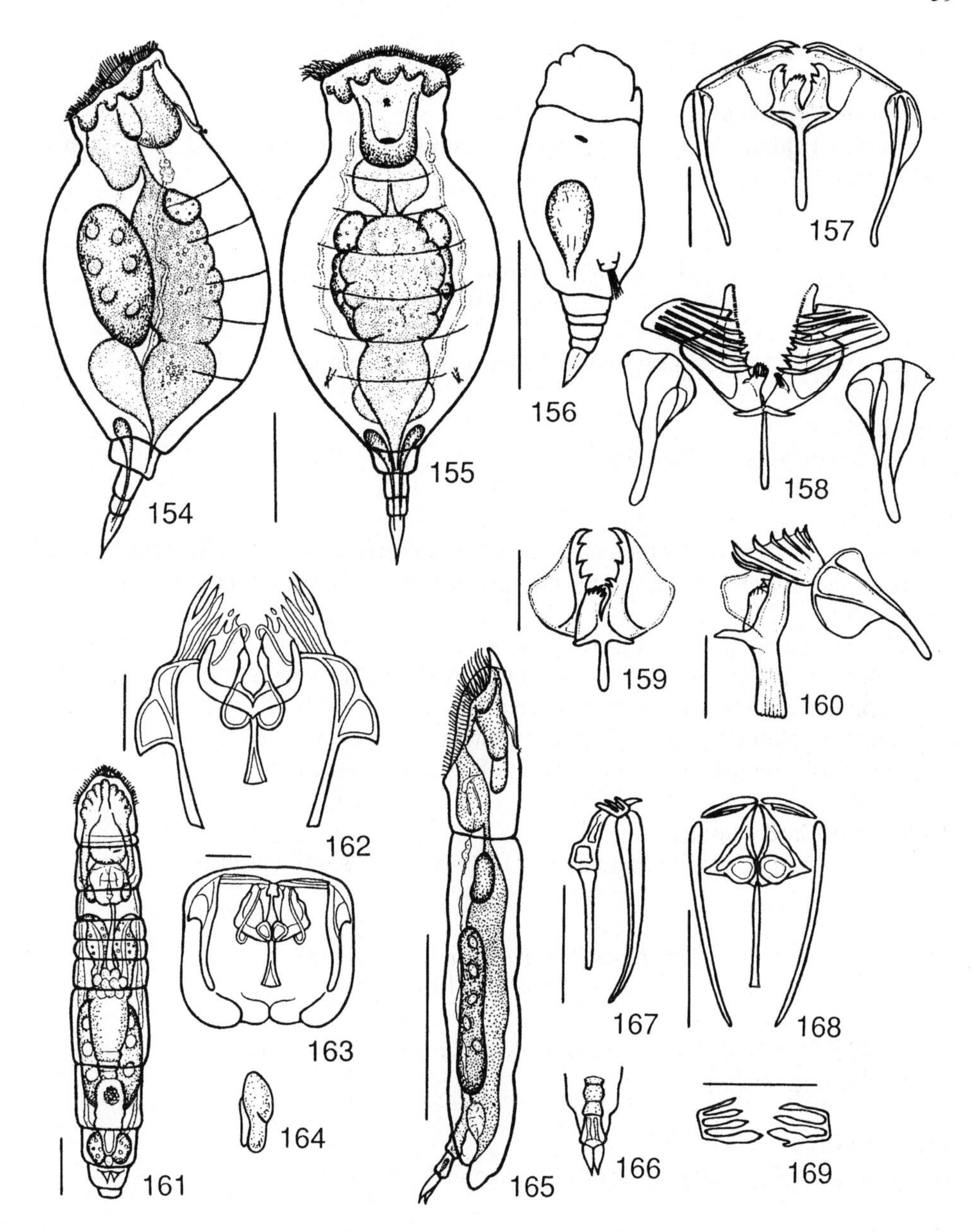

Figs. 154-160: *Proales doliaris*. 154: lateral view, 155: dorsal view, 156: male, dorsal view, 157: trophi, ventral view, 158: trophi, ventro-apical view, 159: incus, apical view, 160: trophi, lateral view.
Figs. 161-164: *Proales cryptopus*. 161: dorsal view, 162: trophi, 163: mastax, ventral view, 164: brain, lateral view.
Figs. 165-169: *Proales macrura*. 165: lateral view, 166: foot and toes, ventral view, 167: trophi, lateral view, 168: trophi, ventral view, 169: unci.
Scale bars: habitus 50 μm, trophi 10 μm.
(154,155,157,159,160: after Harring & Myers, 1924; 156: after Martin, 1977; 158: Hohen Tauern, Austria, leg. et det. C. Jersabek; 161-164: after Wulfert, 1935; 165-169: after Myers, 1933a)

Length 132-145 μm, toe 12-15 μm; trophi 25 μm.

Distribution and ecology
N. America (Maine, New Jersey, Pennsylvania). Among submerged vegetation in acid waters.

19. *Proales indirae* Wulfert, 1966
Figs. 180-183

Wulfert 1966: 86, figs. 44a-d; Koste 1978: pl. 92 figs. 6a-d.

Type locality
Ajwa-reservoir near Baroda, India.

Description
Body fusiform, stout (preserved specimens). Head offset by neck-fold. Trunk elliptical. Foot short, ≤ 1/7 total length, one pseudosegment. Toes ± parallel-sided, tips acute, short.

Rami triangular, inner margins of anterior half with 5-6 curved teeth; points of alulae indicated. Fulcrum relatively short, rod-shaped in ventral view. Unci with 6 teeth, gradually decreasing in size; principal tooth with subsidiary tooth underneath tip. Manubria club-shaped, slightly bent, lateral lamellae medium large.

Length 210 μm, toe 28-31 μm; fulcrum 8 μm, manubrium 20 μm.

Distribution and ecology
India (Baroda city), details on sampling locality and biology not given.

20. *Proales baradlana* Varga, 1958
Figs. 184-189

Varga 1958: 435-440, figs. 8a-c, 9a-d; Koste 1978: 277, pl. 90 figs. 3a-g; Kutikova 1970: 494, figs. 713a-g.

Type locality
Pool near Vaskapu, Baradla Cave near Aggtelek, Hungary.

Description
Body fusiform, stout; cuticle thin, flexible, covered with sand-grains; transparent. Head short, offset by distinct neck-fold. Trunk with 3 pseudosegments. Foot distinctly offset, short c. 1/8-1/10 total length, one pseudosegment. Toes short, broad, tips blunt, slightly curved ventrally. Corona ± frontal, cilia short; laterally two areas with long cilia. No constriction between stomach and intestine. Gastric glands relatively large, rounded. Pedal glands elongated, somewhat irregular. Brain large, ± triangular. Retrocerebral sac long, posteriorly slightly expanded and somewhat dark. Eyespot absent.

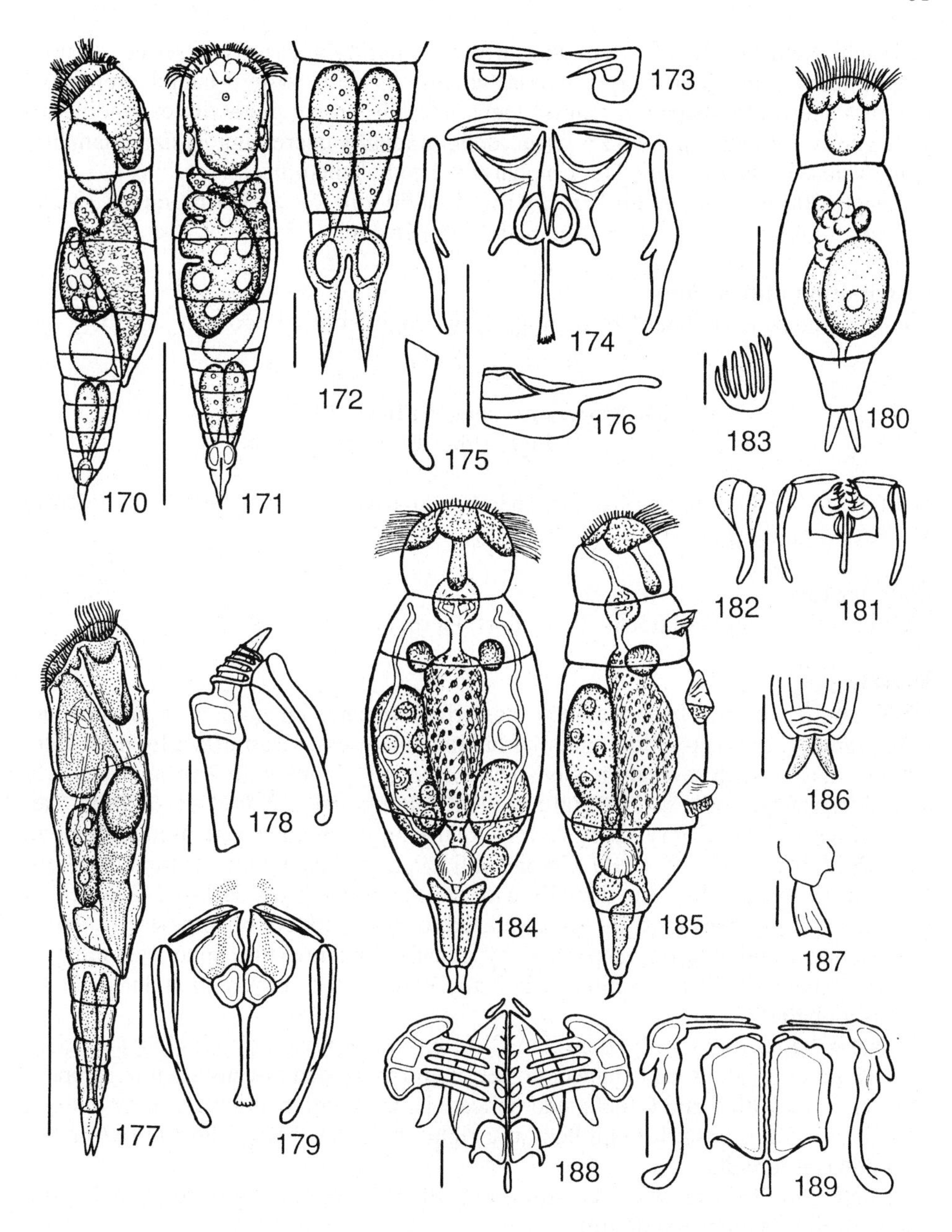

Figs. 170-176: *Proales alba*. 170: lateral view, 171: ventral view, 172: foot and toes, 173: unci, 174: trophi, ventro-apical view, 175: fulcrum, lateral view, 176: manubrium.
Figs. 177-179: *Proales bemata*. 177: lateral view, 178: trophi, lateral view, 179: trophi, ventral view.
Figs. 180-183: *Proales indirae*. 180: ventral view, 181: trophi, 182: manubrium, 183: uncus.
Figs. 184-189: *Proales baradlana*. 184: dorsal view, 185: lateral view, 186: toes, 187: fulcrum, lateral view, 188: trophi, apical view, 189: trophi, dorsal view.
Scale bars: habitus 50 μm (172: 10 μm), trophi 10 μm.
(170-176: after Wulfert, 1939; 177-179: after Myers, 1933a; 180-183: after Wulfert, 1966; 184-189: after Varga, 1958)

Trophi malleate. Rami ± triangular, acutely tapering anteriorly, a series of knob-shaped teeth near inner margins; basal apophyses with postero-lateral projection. Fulcrum short, rod-shaped in ventral view, laterally broad, gradually expanding towards posterior end. Unci with 4 linear teeth, gradually decreasing in size. Manubria long, head with lateral lamellae. Epipharynx 2 rod-shaped pieces.

Length 500-620 μm, width 155-170 μm, foot 60 μm, toe 30-32 μm; ramus 32 μm, fulcrum 9-10 μm, uncus 24-25 μm, manubrium 26 μm.

Distribution and ecology
Europe (Hungary, Baradla Cave). Apparently troglobiont, in pool.

21. *Proales provida* Wulfert, 1938
Figs. 190-193

Wulfert 1938: 389-390, figs. 162a,b, 163a,b; Kutikova 1970: 500, figs. 714a-d; Koste 1978: 279, pl. 93 figs. 3a-d.

Type locality
Limnocrenic spring Goldlocks, Eisersdorf, Germany.

Description
Body stout, fusiform, head rounded, posterior truncate; integument thick, opaque, with granules; greyish-yellow, internal organs colourless. Head offset by neck-fold, anterior margin with several incisions when creeping. Trunk with 7 strong, regularly placed transverse folds and several irregular, longitudinal and transverse folds. Tail small. Foot retracted, very short, c. 1/11 total length, rounded, one pseudosegment. Toes short, bases inflated, acute. Corona oblique, dorsal cilia long, ventral cilia short. Brain large, saccate, hyaline, dorsally with dark granules. Eyes dark-red, 2 large and several small pigment spots (= eye displaced to right with cut off pigment ?), cervical. Stomach and intestine separated by constriction; intestine spherical, ciliated. Gastric glands very large, rounded, laterally compressed, darkly granulated. Pedal glands globular ?

Mastax large. Rami with lateral lamellae; basal apophyses small, elongate, triangular. Fulcrum ± short, rod-shaped, posterior end expanded in ventral view; laterally broader. Unci with 5 and 6-teeth, and delicate lamella with denticulate inner margin. Manubria ± long, posterior end bent at a right angle, head with short posterior and long anterior lamella.

Length 190-200 μm, width 80 μm, toe 8-10 μm; trophi 24 μm: fulcrum 11 μm, uncus 14 μm, manubrium 20 μm.

Distribution and ecology
Europe (Czechia, Poland, Austria). Benthic in calcium rich limnocrenic spring, January, pH 7.5-8.5, among *Fontinalis* ; in littoral of waterway, September, 16°C.

22. *Proales gladia* Myers, 1933
Figs. 194-196

Myers 1933a: 22-23, figs. 13a-c.

Type locality
Jordan Mountain Pond (c. 1100 m a.s.l.), between summits of Sargent and Jordan mountains, Mount Desert Island, Maine, U.S.A.

Description
Body roughly cylindrical, elongate; cuticle thin, sticky, covered with fine particles of detritus. Head not offset by neck-fold. Foot short, c. 1/10 total length, one pseudosegment. Toes slender, undulate, tips acute, slightly recurved. Corona nearly ventral, dorsal arc of circumapical band absent, 2 lateral areas of long cilia. Brain large. Retrocerebral sac and duct clear. Eyespot small, round, near posterior end of brain. Oesophagus extremely short, virtually confluent with stomach; stomach and intestine separated by deep constriction. Gastric glands kidney-shaped. Pedal glands elongate pyriform, foot length.

Trophi modified malleate, asymmetrical. Rami with prominent basal apophyses; alulae unequally developed: right short and blunt, left long, acutely pointed; inner margin of right ramus with blunt tooth near mid-length, left ramus with 3 short teeth. Fulcrum short, rod-shaped in ventral view; laterally plate-shaped. Right uncus with 3, left uncus with 2 clubbed teeth. Manubria unequal, left much longer; posterior ends incurved, knobbed.

Length 150 μm, toe 15 μm; trophi 20 μm.

Distribution and ecology
N. America (Maine), Europe (Sweden). Among submerged *Sphagnum* in acid ponds.

23. *Proales simplex* Wang, 1961
Figs. 203-205

Wang 1961: 161, pl. 15 figs. 140a-c; Kutikova 1970: 496, figs. 706a-c.

Type locality
China.

Description
Body elongate, ± vermiform, tapering towards toes; cuticle soft, flexible. Head offset by neck-fold, with rostrum-like prominence. Corona strongly oblique, 2 lateral ciliary tufts. Posterior 1/3 of trunk often with indistinct transverse folds. Foot short, c. 1/9 total length, broad, one pseudosegment. Toes short, conical, points acute. Eyespot small, posterior to brain, slightly displaced to right. Retrocerebral sac well-developed.

Rami outline oval, median opening ± rhomboid, bases broad, each ramus terminating in 2 appressed teeth. Fulcrum short, in ventral view rod-shaped, knobbed posteriorly. Unci long, single-toothed, curved, tips clubbed, with broad, serrate lamellar plate. Manubria rod-shaped, slightly curved, with lamella. Epipharynx 2 Y-shaped pieces.

Length 120 μm, toe 9 μm; trophi 33 μm.

Distribution and ecology
China. In shallow waters among *Scirpus*.

24. *Proales segnis* Myers, 1938
Figs. 197-202

Myers 1938: 3,5, figs. 1,4,7,11,15,16.

Type locality
Among *Utricularia*, Parvin State Park, Cumberland County, New Jersey, U.S.A.

Description
Body fusiform, swollen, very stout; cuticle soft, flexible, outline variable. Head relatively small, offset by indistinct neck-fold. Corona oblique, small, 2 lateral ciliary tufts. Trunk ± oval, very large. Tail small. Foot small, c. 1/13 total length, stout, one pseudosegment. Toes very short, triangular, tips truncate in lateral view, in dorsal view with swollen bases, then diminishing gradually, ending in tubular points. Eyespot minute, on posterior of relatively small brain. No retrocerebral organ.

Gastric glands very large, oval, Stomach and intestine not separated by constriction; a glandular caecum attached to the ventral wall posteriorly. Pedal glands short, pyriform, completely filling foot.

Trophi modified malleate. Rami broad, triangular, inner margins with prominent, opposing tooth in anterior half; basal apophyses large. Fulcrum ramus length, in ventral view rod-shaped, posteriorly knobbed, laterally broader, slightly tapering posteriorly. Unci with 4 teeth and slender, curved rib resting against dorsal tooth, tips clubbed, gradually decreasing in size. Manubria slightly S-shaped, posterior lamella manubrium length, anterior lamella short.

Length 1050 μm, toe 50 μm; trophi 55 μm.

Distribution and ecology
N. America (New Jersey), Europe (Sweden). Among *Utricularia*, bottom dweller.

25. *Proales phaeopis* Myers, 1933
Figs. 206-209

Myers 1933a: 14-16, figs. 8a-d.

Type locality
Mount Desert Island, Maine, U.S.A.

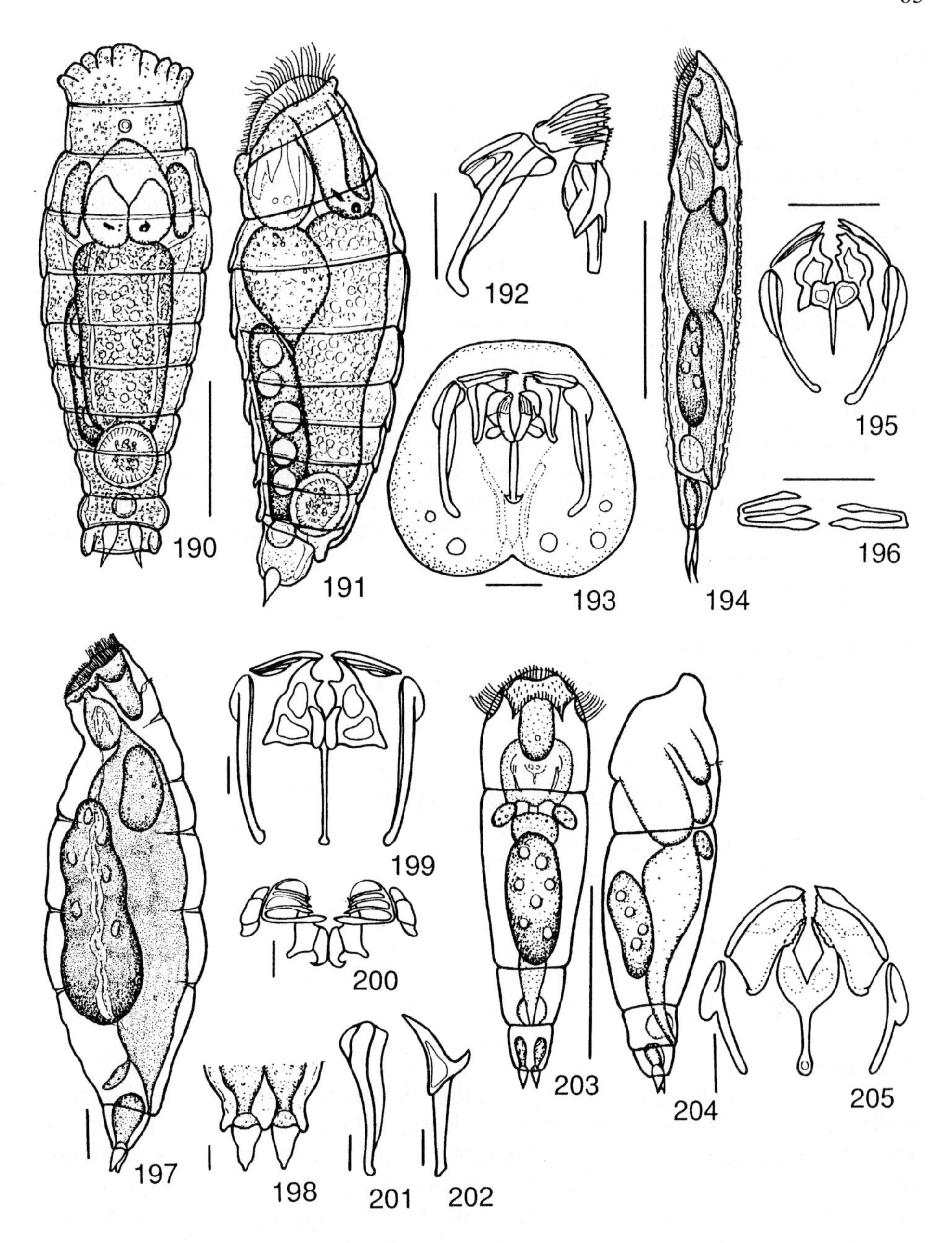

Figs. 190-193: *Proales provida*. 190: dorsal view, 191: lateral view, 192: trophi, lateral view, 193: mastax, dorsal view.
Figs. 194-196: *Proales gladia*. 194: lateral view, 195: trophi, ventral view, 196: unci.
Figs. 197-202. *Proales segnis*. 197: lateral view, 198: toes, dorsal view, 199: trophi, ventral view, 200: trophi, apical view, 201: manubrium, 202: incus, lateral view.
Figs. 203-205: *Proales simplex*. 203: dorsal view, 204: lateral view, 205: trophi, ventral view.
Scale bars: habitus 50 μm, trophi 10 μm.
(190-193: after Wulfert, 1938; 194-196: after Myers, 1933a; 197-202: after Myers, 1938; 203-205: after Wang, 1961)

Description

Body fusiform, elongate, slender; cuticle soft, flexible; very transparent. Head offset by neck-fold. Trunk widest at 1/5 total length, gradually tapering towards toes; posterior half with several dorsal, shallow, transverse folds. Foot short, ≤ 1/15 total length, broad, one pseudosegment. Toes short, conical, ± abruptly reduced to fairly blunt, outcurved tips. Corona nearly frontal, ciliation short, dense; 2 lateral ciliary tufts. Brain large, saccate. Retrocerebral sac absent. A main and secondary eyespot somewhat posteriorly on brain; main eye displaced to right, secondary eye below main one and closer to median line. Stomach not sharply marked-off from intestine. Gastric glands small, kidney-shaped. Pedal glands relatively long, extending into distal pseudosegment of trunk.

Trophi modified malleate. Rami broadest in the middle, right inner margin with 2 stout teeth, left a comb of 5 appressed teeth; alulae absent; basal apophyses rounded, prominent. Fulcrum rod-shaped in ventral view, ending in a fan-shaped expansion; laterally broader, tapering distally. Unci with 4 (5?) clubbed teeth, gradually decreasing in size, attached to subcircular basal plate. Manubria slender, posterior end incurved, knobbed, 2 lateral lamellae 1/2 manubrium length. Epipharynx 2 irregular wedge-shaped pieces.

Length 340 μm, toe 20 μm.

Distribution and ecology

N. America (Maine). Among submerged vegetation in small bodies of acid waters.

26. *Proales pugio* Nogrady, 1983

Figs. 210-214

Nogrady 1983: 111-112, figs. 5a-f.

Type locality and type

Lake Hodges, San Diego County, California, U.S.A. Holotype in Natural Museum of Natural Sciences, Ottawa, Canada, NMCIC 1981-577.

Description

Body uniformly tapering towards toes; cuticle very soft; colourless. Head offset by shallow neck-fold; a small rostrum in swimming position. Foot indistinct, broad, with dorsal papilla between toes. Toes large, broad, dagger-like, tapering to a blunt point, slightly bent towards ventral side, motile, in swimming position at a right angle to the curved body. Corona a large buccal area with weak trochus. Eyespot red, median, on dorso-posterior edge of brain. Retrocerebral organ absent. Stomach and intestine well separated. Gastric glands small, rounded, clear. Pedal glands small, pyriform.

Trophi small, malleate. Rami very small, triangular. Fulcrum in ventral view rod-shaped, distal end slightly expanded; laterally broader. Unci with 6 large teeth and probably some very small ones. Manubria slightly incurved distally; a small, single, ± triangular lamella.

Length 90-120 μm, corona diameter 35-40 μm, toe 15-20 μm; trophi 12x15 μm: uncus 10 μm.

Distribution and ecology
N. America (California), Africa (Kenya). Among shore plants; 20-23 °C, pH 8.0-9.2, 170-1200 μScm^{-1}.

27. *Proales palimmeka* Myers, 1940
Figs. 215-219

Myers 1940: 6-7, pl. 1 figs. 6-7, pl. 2 figs. 15, 16, 19.

Type locality
Promised Land Lake, Pike County, Pocono Plateau, Pennsylvania, U.S.A.

Description
Body ± fusiform, slender; hyaline. Head small, offset by dorsal transverse fold. Trunk decreasing abruptly in diameter at half-length, continuing cylindrically to foot. Tail distinct. Foot short, c. 1/11 total length; one pseudosegment, postero-dorsal margin with dorsally projecting spur between toes. Toes short, conical, bases somewhat swollen, acutely pointed. Corona oblique, 2 lateral areas with long cilia. Brain long, saccate. Retrocerebral sac and duct rudimentary, confluent with brain. Eyespot rudimentary, scattered pigment granules, placed to the right. No constriction between stomach and intestine. Gastric glands small, oval. Pedal glands pyriform, short, almost confined to foot.

Trophi modified malleate, relatively small. Rami lyrate, inner margins without teeth, outer margins with forwardly projecting blunt tooth; basal apophyses large, oval. Fulcrum long, in ventral view slender, rod-shaped, posterior end slightly expanded; laterally broader, dorsal margin straight, ventral margin curved. Unci with 5 well-developed, gradually diminishing teeth. Manubria long, ± straight, with small posterior and anterior lamellae.

Length 210 μm, toe 12 μm; trophi 34 μm.

Distribution and ecology
N. America (Pennsylvania). Among submerged *Sphagnum* and algae in lakes; pH 6.4-6.5.

28. *Proales cognita* Myers, 1940
Figs. 220-224

Myers 1940: 5-6, pl. 1 figs. 3-4, pl. 2 figs. 6, 13, 14.

Type locality
Pocono and Naomi Lakes, Monroe County, Pocono Plateau, Pennsylvania, U.S.A.

Description
Body stout fusiform, hyaline, milky. Head not offset by neck-fold; neck region well marked. Distal half of trunk with dorsal transverse fold. Foot short, c. 1/8 total length, broad, one pseudosegment; dorsally between toes a bulbous process surmounted by

papilla; posteriorly 2 latero-dorsally directed, cuticular projections. Toes short, conical, somewhat expanded near base, tapering to small acute tips. Corona oblique, 2 lateral areas with long cilia. Brain large, saccate. Retrocerebral sac rudimentary, confluent with dorsal portion of brain. Eyespot rudimentary; scattered pigment granules mid-length on brain, displaced to right. No constriction between stomach and intestine. Gastric glands relatively small, kidney-shaped. Pedal glands very large, pyriform, with prominent mucus reservoir at base of each toe.

Trophi modified malleate. Rami triangular; inner margin of left ramus with 2 blunt teeth, right ramus with single blunt tooth near mid-length; basal apophyses large, oval. Fulcrum long, slender, rod-shaped in ventral view, posterior end slightly expanded.

Unci 6-toothed, a single, bifurcate principal tooth and 5 ± rudimentary teeth. Manubria long, ± straight, posterior lamella manubrium length, anterior one small.

Length 245 μm, toe 18 μm; trophi 38 μm.

Distribution and ecology
N. America (Pennsylvania). Among *Nitella* and *Sphagnum cuspidatum* Ehrh. in lakes; pH 6.2-6.4.

29. *Proales gigantea* (Glascott, 1893)
Figs. 225-233

Synonym: ? *P. ovicola* Giard, 1908

P. daphnicola after Bartoš (1959), Rudescu (1960).

Glascott 1893: 80-82, pl. 7 fig. 1 (*Notommata gigantea*); ? Giard 1908: 184 (*Proales ovicola*); Stevens 1912: 481, pl. 24 figs. 1-5 (*Proales gigantea*); Budde 1924: 754; Harring & Myers 1924: 424-425, pl. 17 figs. 6-10; Kutikova 1970: 498, figs. 712a-d; Koste 1972: 153, pl. 2 figs. 1a-g, text figs. 1,2; Koste 1978: 282, fig. 26:2, pl. 86 figs. 6a,b, pl. 91 figs. 1a-g, 2a-d; Koste & Shiel 1990: 135, figs. 2.4a-e.

Type locality
Wexford, Ireland.

Description
The full-grown, parasitizing female takes the form of a shapeless, distended bag. Body of free-swimming individuals squat fusiform, cylindrical; cuticle very soft, flexible, outline constantly changing. Head short, broad. Neck represented by 2 or 3 indistinct dorsal transverse folds. Trunk distally with some dorsal, transverse folds. Tail distinct, less prominent. Foot short, ≤ 1/6 total length, broad, apparently with 2 pseudosegments by the presence of a dorsal transverse fold; terminating in a hemispherical bulb with 2 toes set far apart. Toes short, broad conical, abruptly reduced to short, blunt points. A prominent, acutely pointed and slightly curved spur dorsally between toes. Corona oblique, weakly ciliated, 2 lateral tufts of long cilia. Stomach and intestine not separated by constriction. Gastric glands small, ± globular. Pedal glands large, elongated fusiform, with reservoirs. Brain rather small, saccate; a small

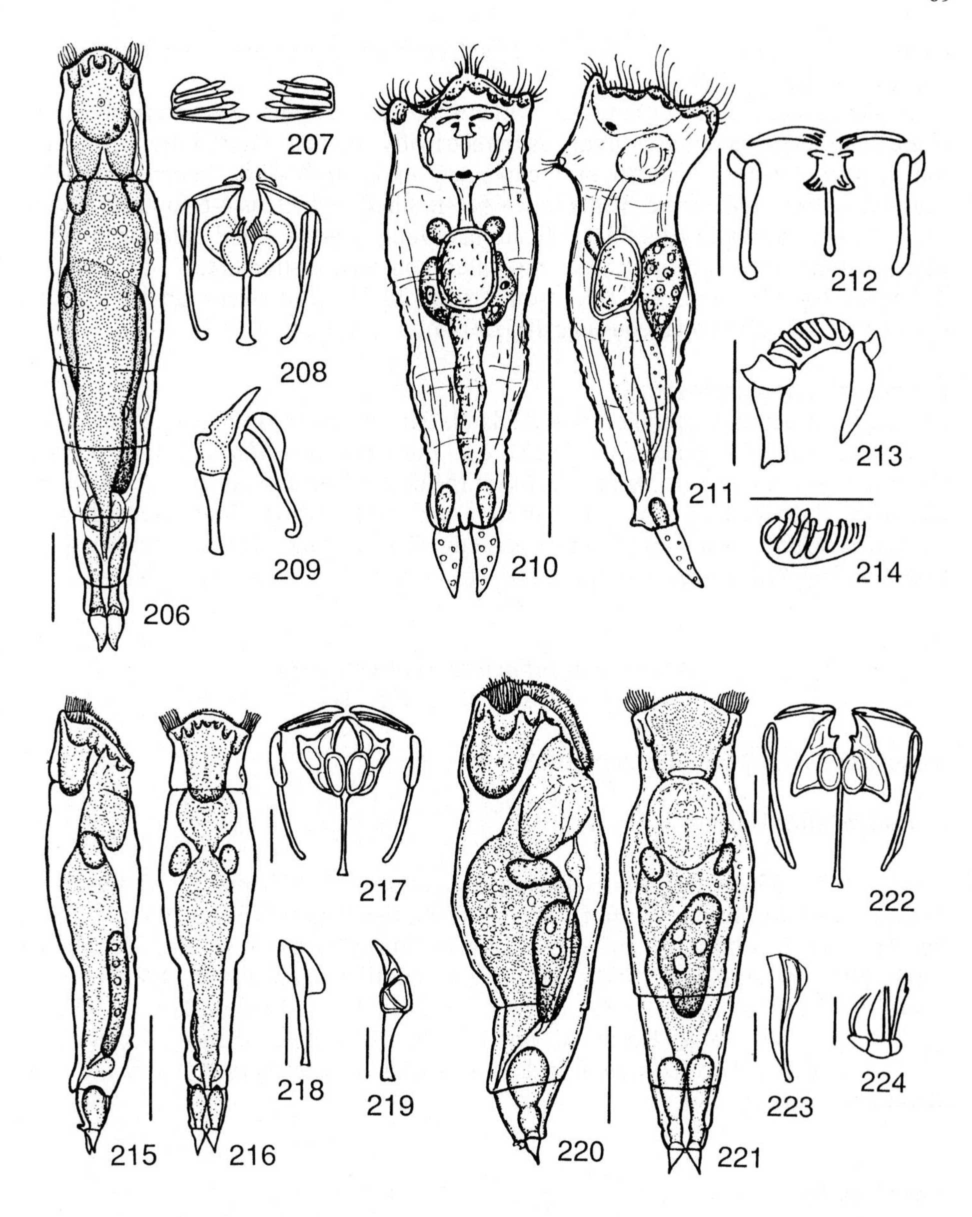

Figs. 206-209: *Proales phaeopis*. 206: dorsal view, 207: unci, 208: trophi, ventral view, 209: incus and manubrium, lateral view.
Figs. 210-214: *Proales pugio*. 210: dorsal view, 211: lateral view, 212: trophi, dorsal view, 213: trophi, lateral view, 214: uncus.
Figs. 215-219: *Proales palimmeka*. 215: lateral view, 216: dorsal view, 217: trophi, ventral view, 218: manubrium, 219: incus, lateral view.
Figs. 220-224: *Proales cognita*. 220: lateral view, 221: ventral view, 222: trophi, ventral view, 223: manubrium, 224: uncus.
Scale bars: habitus 50 μm, trophi 10 μm.
(206-209: after Myers, 1933a; 210-214: after Nogrady, 1983; 215-224: after Myers, 1940)

red eyespot at posterior end, eye with crystalline body, sometimes displaced to right; eye less visible in adults.

Trophi malleate. Rami triangular, without alulae; inner margins anteriorly finely denticulate; basal apophyses large, asymmetrically toothed. Fulcrum long, slender, rod-shaped, posterior end slightly expanded in ventral view, laterally broader, slightly tapering and curved. Right uncus 5-6-toothed, left uncus 4-6-toothed; principal teeth with accessory toothlet. Manubria ± long, posterior and anterior lamellae short, in anterior half. Epipharynx 2 irregular, conchoidal structures.

Length 140-510 μm, toe 8-12 μm; trophi 30-35 μm: uncus up to 19 μm, manubrium 18 μm; subitaneous egg 105x50 μm.

Distribution and ecology

Europe, N. America, Australia, New Zealand. Parasitic in eggs of several freshwater snails (Stevens 1912; Nekrassow 1928), e.g. *Lymnaea auricularia* (L.), *L. peregra* (Müll.), *L. stagnalis* (L.), *Myxas glutinosa* (Müll.), *Physa fontinalis* (L.), *Viviparus viviparus* (L.); feeding on snail embryo and surrounding fluid. According to Koreneva (1958) also ovipositing in egg masses and feeding on eggs of Chironomidae (*Limnochironomus*, *Endochironomus*, *Cryptochironomus*).

30. *Proales fallaciosa* Wulfert, 1937

Figs. 234-243, pl. 12 figs. 1-4

Synonym: ? *P. tyrphosa* Bērziņš, 1948

P. sordida after Harring & Myers, 1922

Wulfert 1937: 65-66, figs. 4a-f; Harring & Myers 1922: 605-606, pl. 51 figs. 9-12 (*Proales sordida* Gosse); Wulfert 1939: 597-599, figs. 12a-i; Wulfert 1940: 583-585, fig. 23; Donner 1955: 102-103, figs 31a-c; ? Wulfert 1956: 483, fig. 35; Donner 1964: 296, fig. 34; Koste 1968a: 149, fig. 30; Kutikova 1970: 496, 498, figs. a-f; Koste 1976: 210, pl. 21 figs. 4a-d; Koste 1978: 281-282, pl. 86 figs. 2a-d, pl. 92 figs. 3a-d, 4a,b, 5a-i, 6a-g; Koste & Shiel 1990: 135, figs. 2.3a-d; Jersabek & Schabetsberger 1992a: 97, figs. 37a-d; Jersabek & Schabetsberger 1992b: 71, figs. 30a-c.

Type locality

Netzschkauer Schachtteich, near the Chaussee Merseburg, Lauchstädt, Germany.

Description

Body roughly cylindrical to fusiform, elongate, slender; cuticle soft, flexible, outline variable; colourless, hyaline. Head and neck offset by transverse folds. Trunk with pseudosegments, usually with dorsal, longitudinal folds. Tail less distinct. Foot short, ≤ 1/8 total length, broad, 2 pseudosegments; small rounded knob projecting dorsally between toes. Toes short, conical, tips ± abruptly reduced to acute points. Corona slightly oblique, laterally 2 strongly ciliated auricle-like tufts. Brain large, saccate. Eyespot small, disc-shaped, red, on brain. Eyespot and hemispherical retrocerebral sac displaced to right. Stomach and intestine separated by marked constriction. Gastric glands globular, elongate or slightly 3-lobate. Pedal glands variable, elongate, pyriform, footh length, with mucus reservoir.

Trophi virgate. Rami truncate triangular in dorsal view, with plate oriented towards median plane of incus; tips with 2 blunt teeth; inner margins with shallow indentations anteriorly and large, 1-2-pointed tooth medially; basal apophyses large, ± hemispherical with broad, short projection; dorsal fenestrae central. Fulcrum with mid-ventral crest, in ventral view gradually expanding towards ± broad, fan-shaped, scalloped end; in lateral view with broad base, tapering, posterior end slightly expanded. Unci with 4-7 acute teeth, gradually decreasing in size, principal teeth bifurcate; inner side with plate covered by skleropili near anterior margin. Manubria with short posterior and long, broad anterior cavity almost extending till slightly curved manubrium end; posterior cavity with opening on ventral margin; epipharynx 2 asymmetrical, irregular, club-shaped pieces.

Subitaneous egg elongate oval (fig. 243).

Length 200-320 μm, toe 9-15 μm; trophi 25-28 μm: ramus 9-12 μm, fulcrum 6-11 μm, uncus 9-17 μm; manubrium 17-22 μm, epipharyngeal element 6-11 μm; subitaneous egg 56-69x30-36 μm.

Distribution and ecology

Cosmopolitan. In alkaline to slightly acid waters, in submerged mosses and among decomposing macrophytes. Perennial. Food: detritus, bacteria, algae, decomposing microcrustaceans and macroinvertebrates. Cultering, food, etc.: Pourriot (1965); life-history: Jennings & Lynch (1928, sub *P. sordida*), Liebers (1937, sub *P. sordida*), Dorazio (1984, sub *P. sordida*), Pourriot (1965); *P.* cf. *fallaciosa*: Koste & Shiel, 1986.

31. *Proales ornata* Myers, 1933

Figs. 244-248

Myers 1933a: 18-19, figs. 10a-e.

Type locality

Mount Desert Island, Maine, U.S.A.

Description

Body fusiform, elongate, slender; cuticle soft, flexible, outline quite constant. Head distinctly offset by neck-fold, relatively long, pointed; apex a bare cuticular area limited posteriorly by fold of integument. Trunk cylindrical, gradually tapering to foot, distally 2 transverse folds. Tail semi-circular, distinct. Foot moderately long, c. 1/5 total length; 2 pseudosegments, terminal twice length basal, its posterior edge with 2 dorso-lateral, spur-like, cuticular processes. A dorsal, bulbous process surmounted by minute papilla between toes. Toes short, stout, inflated at base, abruptly reduced to short, blunt tips. Corona very oblique, 2 lateral ciliary areas. Brain long. No indication of retrocerebral sac. Eye disc-shaped, red, with lens, on brain, displaced to right. Stomach not separated from intestine. Gastric glands kidney-shaped. Pedal glands elongate, extending till mid-length basal foot pseudosegment.

Trophi modified malleate. Rami triangular, anterior ends pointed, inner margins

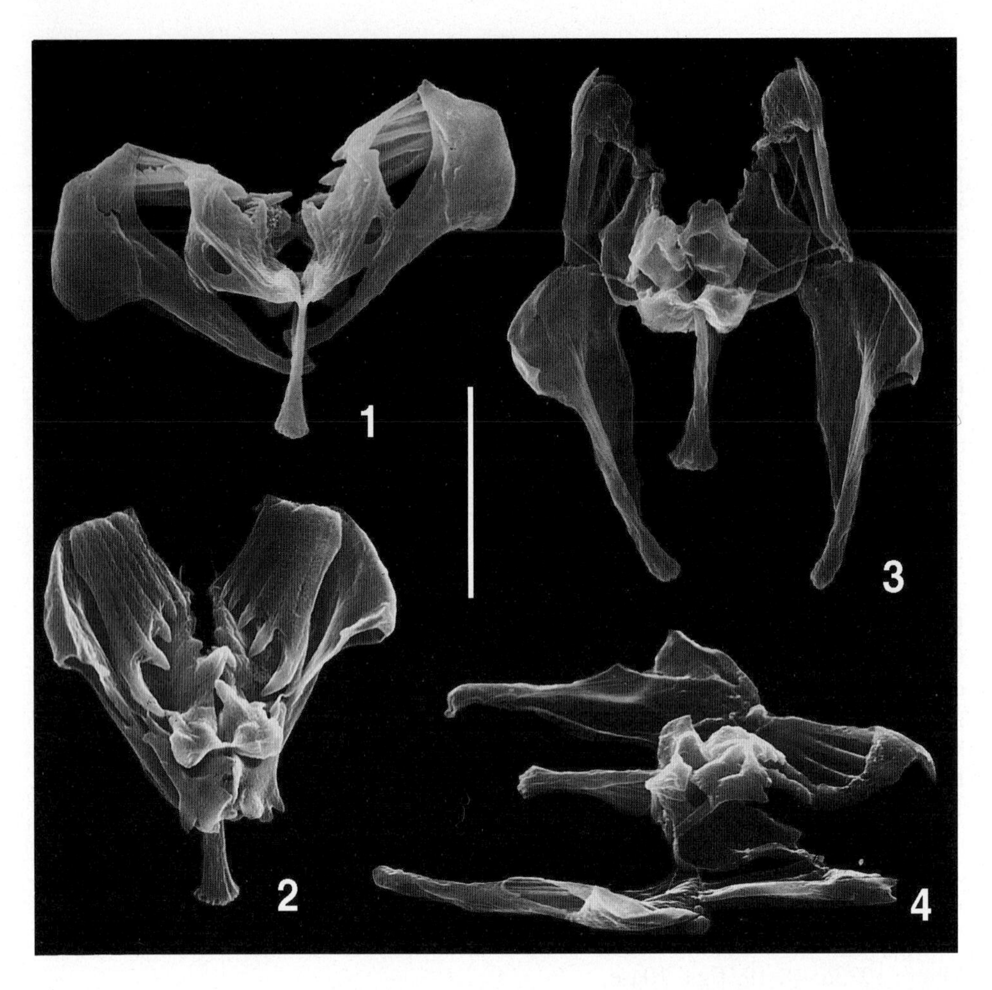

Plate 12. *Proales fallaciosa*, trophi (S.E.M. photographs). 1: dorsal view, 2,3 ventral view, 4: lateral view.
Scale bar: 10μm.
(1: Blankaart, W. Vlaanderen, Belgium; 2,4: Brilschans, Antwerpen, Belgium; 3: Canigou, Pyrenées Orientales, France)

with strong, blunt tooth near mid-length; lateral blunt alulae; basal apophyses large, rounded. Fulcrum ± long, in ventral view rod-shaped, posterior end with fan-shaped expansion; laterally gradually tapering to oblique posterior end. Unci slightly asymmetrical, 6 teeth each, gradually decreasing in size, principal tooth bifurcate. Manubria long, nearly straight, posterior and anterior lamellae 3/4 and about 1/2 manubrium length respectively. Epipharynx 2 wedge-shaped pieces.

Length 254-278 μm, toe 14-18 μm; trophi 35 μm.

Distribution and ecology

N. America (Maine, Pennsylvania), Europe (Sweden). Among submerged *Sphagnum* in acid waters; pH 6.2-6.5.

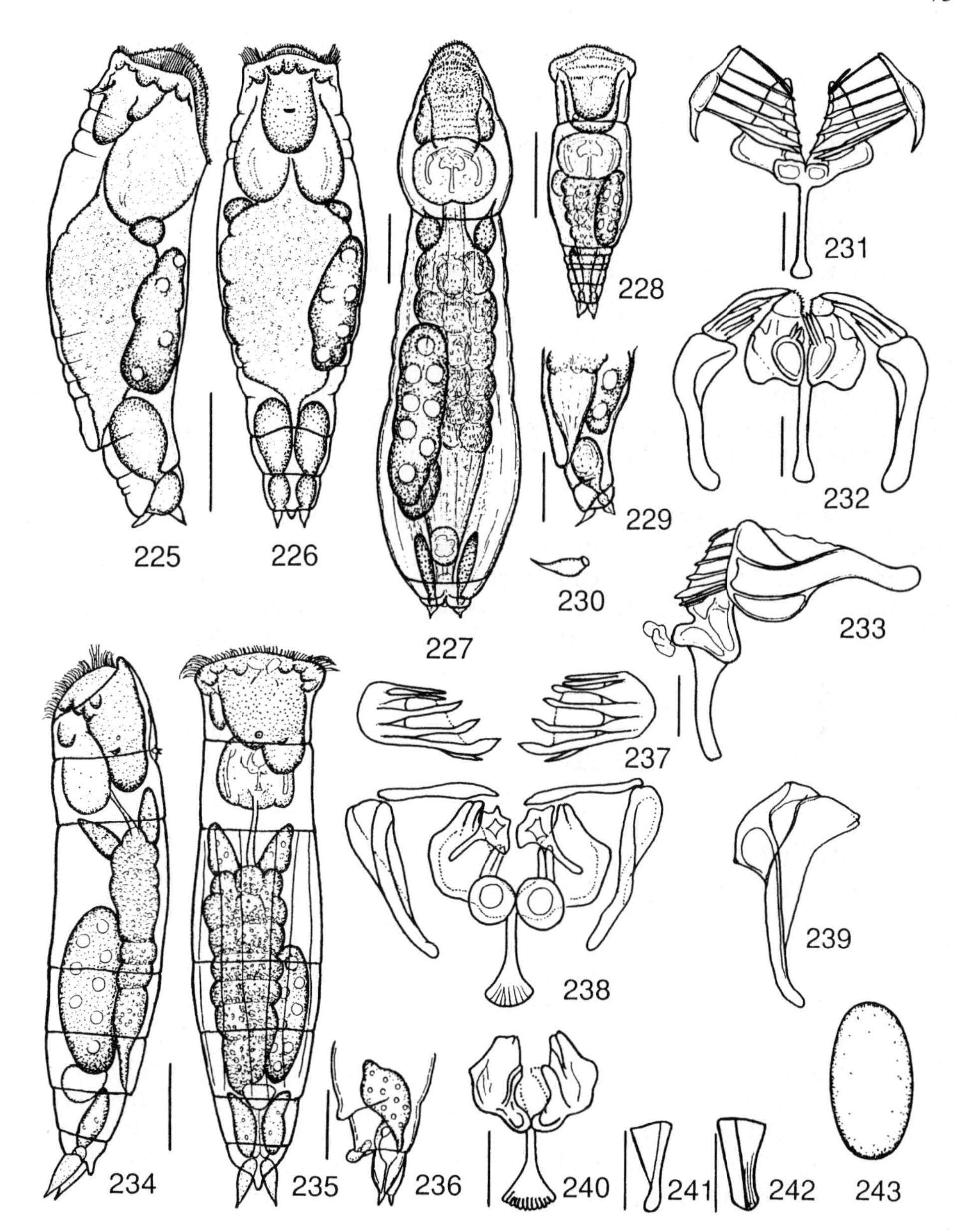

Figs. 225-233: *Proales gigantea*. 225: lateral view, 226: dorsal view, 227: ventral view, 228: juvenile, 229: posterior of body, lateral view, 230: spur, 231: trophi, apical view, 232: trophi, ventral view, 233: trophi, lateral view.

Figs. 234-243: *Proales fallaciosa*. 234: lateral view, 235: dorsal view, 236: posterior of body, lateral view, 237: unci, 238: trophi, ventral view, 239: manubrium, 240: incus, ventral view, 241,242: fulcrum, lateral view, 243: subitaneous egg.

Scale bars: habitus 50 μm (236: 10 μm), trophi 10 μm.

(225,226,231,233: after Harring & Myers, 1924; 227-230,232: after Koste, 1972; 234,235,241: after Wulfert, 1939, 236: after Wulfert, 1956; 240: after Kutikova, 1970, 237-239,242-243: Brilschans, Antwerpen, Belgium)

32. *Proales wesenbergi* Wulfert, 1960
Figs. 249-252

Wulfert 1960: 326-327, figs. 53a-d; Kutikova 1970: 500, figs. 722a-d; Koste 1978: 278-279, pl. 92 figs. 9a-d.

Type locality
Stratiotes pond near Winkelmühle, village pond near Tornau, Dübener Heide, Germany.

Description
Body ± fusiform, width increasing from head towards distal third of trunk, afterwards decreasing. Head and neck offset by distinct transverse folds; neck somewhat longer than head. Trunk with long anterior pseudosegment and 2 short distal ones. Tail broad, lying over basal foot pseudosegment. Foot very short, c. 1/12 whole length of animal, 2 pseudosegments, terminal pseudosegment dorsally with stiff, posteriorly expanding plate. Toes short, spread, inner margins outcurved, acutely pointed. Corona oblique, no lateral ciliary tufts. Eye red, disc-shaped, displaced somewhat to left, at limit between brain and retrocerebral sac. No constriction between stomach and intestine. Gastric glands very small, hyaline. Pedal glands elongate.

Trophi insufficiently described. Rami brown coloured. Fulcrum long, in ventral view slender, rod-shaped with knob-like posterior end. Manubria long, extending beyond fulcrum, broadly incurved.

No measurements.

Distribution and ecology
Europe (Germany). In acid water ponds; autumn.

33. *Proales sordida* Gosse, 1886
Figs. 253-259

Gosse *in* Hudson & Gosse 1886: 37-38, pl. 18 figs. 7,7a-c; Harring 1913 (*Pleurotrocha sordida*); Wulfert 1937: 67, figs. 5a-h; Wulfert 1939: 596-597, figs. 14a-f; Wulfert 1956: 484, figs. 37a-f; Kutikova 1970: 499, figs. 718a-f; Koste 1978: 279-280, pl. 92 figs. 7a-f; Koste & Shiel 1990: 136, figs. 3.3 a-e; *non P. sordida* Harring & Myers 1922: 605-606, pl. 51 figs. 9-12.

Type locality
Not specified; “Many localities in England and Scotland: common in pools”.

Description
Body fusiform; fairly hyaline, colourless. Head and neck pseudosegment distinct; head slightly flared anteriorly. Trunk with delicate, dorsal longitudinal folds. Foot long, c. 1/4, total length, broad, 3 pseudosegments, terminal pseudosegment projecting over bases of toes. Toes short, conical, abruptly reduced to acute, motile points. Corona oblique, 2 weakly defined lateral auricle-like tufts. Eyespot large, on

posterior of brain, displaced to right. Retrocerebral sac large, hemispherical. Older individuals with granular accumulation in anterior part of head. Stomach and intestine not separated by constriction. Gastric glands small, globular. Pedal glands large, foot length, reservoirs present. Brain large, saccate.

Rami with large, acute alulae, rounded basal-apophyses and 2 triangular plates. Fulcrum long in ventral view, slender, rod-shaped with expanded and scalloped posterior end; laterally ± parallel sided. Unci 5-toothed, principal tooth with accessory tooth underneath tip, dorsal tooth linear. Manubria ± long, slightly curved posteriorly, posterior and anterior lamellae long, almost extending till posterior end. Epipharynx 2 irregular, incised platelets with stout, lateral spine.

Length 150-230 μm, toe 10-12 μm; trophi 25 μm: fulcrum 12 μm, uncus 12 μm, manubrium 20 μm.

Distribution and ecology

Cosmopolitan. In periphyton and diatom films.

34. *Proales micropus* (Gosse, 1886)

Figs. 260-266

Gosse *in* Hudson & Gosse 1886: 46, pl. 19 figs. 12,12a (*Furcularia micropus*); Jennings 1901: 743, pl. 5 fig. 82 (*Proales micropus*); Rodewald 1935: 204, figs. 6,6a-c; Rudescu 1960: 684, figs. 560a-c; Koch-Althaus 1963: 431, figs. 42a-c; Kutikova 1970: 496, figs. 720a-e; Koste 1978: 281, pl. 92 figs. 2a-g; Koste & Shiel 1990: 135-136, figs. 2.5a-e.

Type locality

A ditch near Birmingham, England.

Description

Body elongate, slender, fusiform or tapering towards toes; cuticle soft, flexible, outline variable; colourless. Head without distinct neck-fold; rostrum small, may be extended. Trunk with transverse folds. Foot small, c. 1/6-1/10 total length, with 3 pseudosegments. Toes short, broad conical, ± abruptly reduced to short points, often with inner convexity; toes usually directed ventrally. Corona oblique, no lateral ciliary tufts. Eyespot red, on brain, displaced to right, sometimes absent. No constriction between stomach and intestine.

Trophi slightly curved in lateral view. Fulcrum rod-shaped in ventral view, with triangular posterior expansion. Unci 3-toothed (?). Manubria long, 2 small lamellae on outside. Epipharynx 2 small platelets.

Length 100-150 μm, toe 6-9 μm; trophi 14-16 μm.

Distribution and ecology

Europe, N. and S. America, Australia. In periphyton, on *Chara*, in ponds, lakes and moor waters.

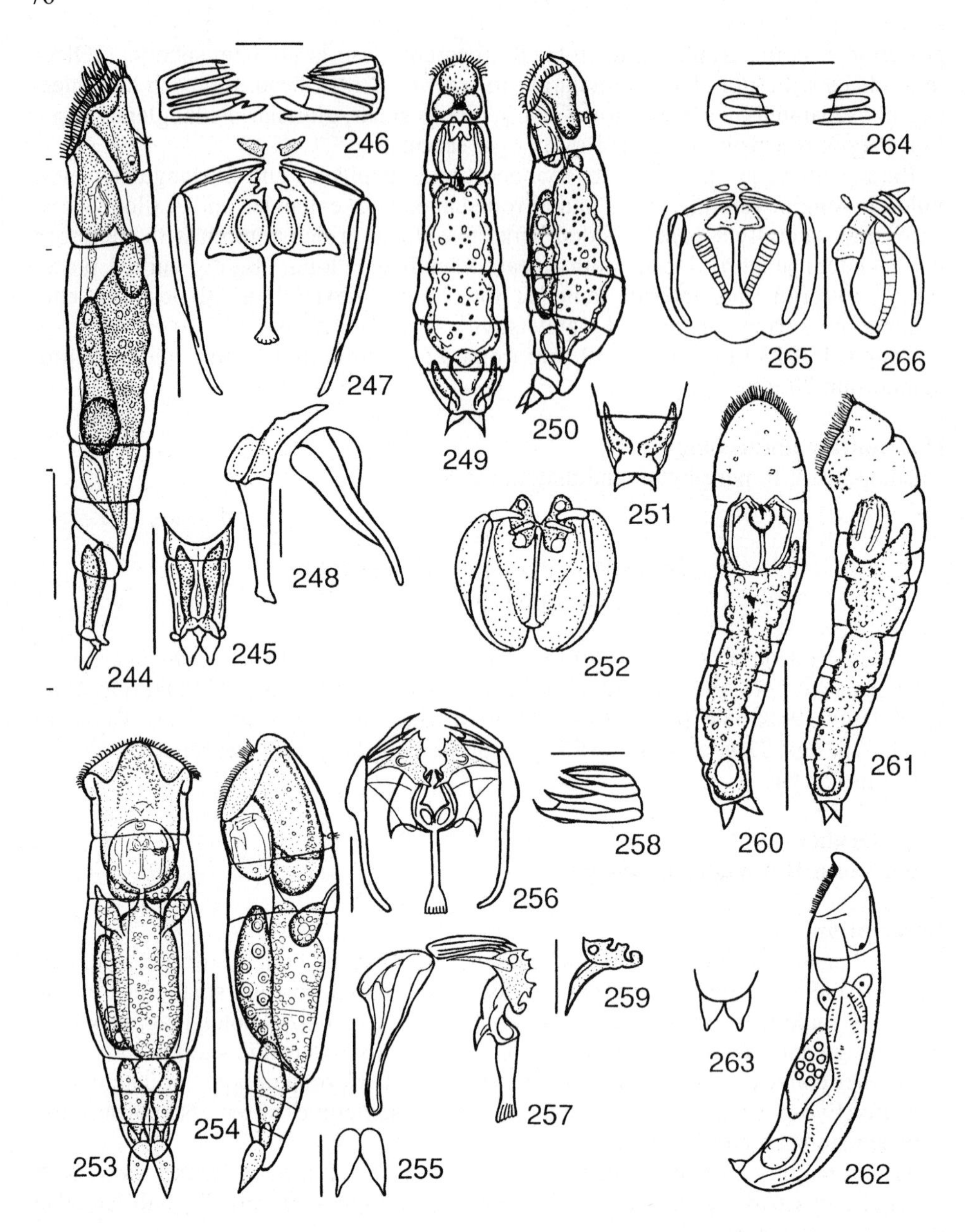

Figs. 244-248: *Proales ornata*. 244: lateral view, 245: foot and toes, dorsal view, 246: unci, 247: trophi, ventral view, 248: incus and manubrium, lateral view.
Figs. 249-252: *Proales wesenbergi*. 249: dorsal view, 250: lateral view, 251: foot, 252: mastax.
Figs. 253-259: *Proales sordida*. 253: dorsal view, 254: lateral view, 255: toes, 256: trophi, ventral view, 257: trophi, lateral view, 258: uncus, 259: epipharyngeal element.
Figs. 260-266: *Proales micropus*. 260: dorsal view, 261: lateral view, 262: lateral view, 263: toes, 264: unci, 265: trophi, ventral view, 266: trophi, lateral view.
Scale bars: habitus 50 μm (255,263: 10 μm), trophi 10 μm. (244-248: after Myers, 1933a; 249-252: after Wulfert, 1960; 253,254,256-259: after Wulfert, 1939; 255: after Wulfert, 1937; 260,261: after Hudson & Gosse, 1886; 262,263: after Koch-Althaus, 1963; 264-266: after Rodewald, 1935).

35. *Proales adenodis* Myers, 1933
Figs. 267-270

Myers 1933a: 16-17, figs. 9a-d.

Type locality
Witch Hole and Long Lake, Mount Desert Island, Maine, U.S.A.

Description
Body fusiform, elongate, slender; cuticle very flexible, outline variable; hyaline. Head distinctly offset by deep neck-fold, narrow. Trunk widest at 1/3 body length, gradually tapering towards toes. Foot fairly long, c. 1/5 total length, 2 pseudosegments, terminal pseudosegment twice length basal. Toes short, stout, bases slightly inflated, ± abruptly tapering to acute tips; a septum divides tips from bases. Corona slightly oblique, 2 lateral areas with long cilia. Brain long, saccate. Retrocerebral sac reduced. A main and secondary eyespot on brain, somewhat posteriorly, displaced to right; secondary eyespot below main one and closer to median line. No constriction between stomach and intestine. Gastric glands very large, kidney-shaped, giving opposite region of trunk a swollen appearance. Pedal glands elongate, foot length, each with a pair of small rudimentary glands attached to their bases.

Trophi small, modified malleate. Rami with large, ± triangular basal apophyses; inner margin of left and right ramus with one and 2 blunt teeth respectively; anterior end of rami pointed. Fulcrum long, stout, rod-shaped in ventral view, ending in fan-shaped expansion; laterally with broad base, ± abruptly narrowing near mid-length. Unci with 5 teeth, gradually decreasing in size. Manubria long, stout, with posterior and anterior lamellae. Epipharynx 2 irregular, ± wedge-shaped pieces.

Length 290 µm, toe 15 µm.

Distribution and ecology
N. America (Maine). Among aquatic vegetation in acid waters.

36. *Proales granulosa* Myers, 1933
Figs. 271-274

Myers 1933a: 19-21, figs. 11a-d.

Type locality
Mount Desert Island, Maine; Atlantic County, New Jersey, U.S.A.

Description
Body fusiform, elongate, slender; outline quite constant. Head small, neck-fold absent. Trunk with lateral swellings due to gastric glands, gradually tapering towards toes; distally 2 transverse folds. Foot short, ≤ 1/8 total length, 2 pseudosegments, posterior portion of terminal pseudosegment somewhat divided, giving toes appearance being twice as long as they actually are. Toes short, conical, acutely pointed. Buccal field covered with short, densely set cilia, 2 lateral areas with long cilia. Brain large,

saccate. Retrocerebral sac rudimentary, with small granules and indications of Y-shaped duct. Eyespot on brain, displaced to right. Stomach separated from intestine by shallow constriction; stomach wall of adult female crowded with minute algae. Gastric glands very large, kidney-shaped, granulose with central clusters of clear vacuoles. Pedal glands elongate, foot-length.

Trophi modified malleate. Rami lyrate, laterally large, thin, semi-lunar plates; prominent basal apophyses. Fulcrum rod-shaped in ventral view, fairly long, posterior end expanded; laterally tapering from broad base to small, fan-shaped posterior expansion. Unci with 2 clubbed teeth each, united by web-like basal plate. Manubria long, slender, curved, posterior and anterior lamellae 1/2 manubrium length. Epipharynx 2 small, irregular, wedge-shaped pieces.

Length 140 μm, toe 6 μm.

Distribution and ecology

N. America (Maine, New Jersey, Pennsylvania). Among submerged *Sphagnum* and aquatic vegetation in acid waters; pH 4.0-6.8.

37. *Proales decipiens* (Ehrenberg, 1832)

Figs. 275-286

Synonym: *P. vermicularis* Dujardin, 1841
P. brevipes Harring & Myers, 1924

Ehrenberg 1832: 132 (*Notommata decipiens*); Dujardin 1841: 648, pl. 21 fig. 7 (*Notommata vermicularis*); Hudson & Gosse 1886: 36, pl. 18 figs. 6,6a (*Proales decipiens*); Von Hofsten 1910 (*Pleurotrocha decipiens*); Harring & Myers 1922: 603-605, pl. 51 figs. 5-8; Harring & Myers 1924: 428-429, pl. 19 fig 1,2 (*Proales brevipes*); Donner 1943: 25-26, figs. 3a-c; ? Wulfert 1956: 483, fig. 35; Wulfert 1961: 94, figs. 29a-i; ? Wulfert 1968: 414, figs. 8a-h; Koste 1968a: 149, fig. 31; Kutikova 1970: 490,492, figs. 705a-d; Koste 1978: 280, pl. 86 figs. 1a-d, pl. 92 figs. 1a-l; Braioni & Gelmini 1983: 89, figs. 42a,b, figs. 43d-i,r-v; Koste & Shiel 1990: 133,135, *non* figs. 2.1a-e.

Type locality

Near Berlin, Germany.

Description

Body roughly cylindrical to fusiform, elongate, slender; cuticle soft, flexible, outline fairly constant; very hyaline. Head offset by neck-fold; rostrum small, distinct. Neck pseudosegment short, slightly wider than head. Trunk with pseudosegments, dorsally longitudinal folds. Foot short, c. 1/7-1/10 total length, broad, 2 pseudosegments. Toes relatively short, stout, conical,acutely pointed; inner margins of bases occasionally slightly inflated. Corona oblique, laterally 2 strongly ciliated tufts. Brain very large, saccate. Retrocerebral sac pyriform, somewhat to right of median line. Eyespot small, red, on brain, usually displaced to right, rarely to left, crystalline bodies present. Stomach and intestine separated by ± deep constriction. Gastric glands rounded, oval or 3-lobate. Pedal glands large, pyriform, foot length.

Trophi virgate. Rami triangular, inner margins and tips asymmetrically toothed; alulae absent; basal apophyses large, bearing asymmetrical teeth. Fulcrum short, rod-shaped with expanded posterior end in ventral view; laterally with broad base, straight ventral margin and tapering dorsal margin, posterior end expanded. Left uncus with 4-5-teeth, right 5 teeth, gradually decreasing in size, principal tooth bifurcate. Manubria ± long, club-shaped, posterior end slightly incurved; posterior lamella long, anterior one somewhat shorter. Epipharynx 2 club-shaped pieces.

Length 120-270 μm, toe 10-16 μm; trophi 15-21 μm.

Male known (fig. 280); numerous in autumn.

Distribution and ecology

Cosmopolitan. Among vegetation in small water bodies, ponds etc.; in wet *Sphagnum*. Food: detritus, bacteria, small algae. The species has also been reported (Stevens, 1907) to feed on adults and developing eggs of *Stephanoceros fimbriatus* (Goldf.) (Rotifera: Collothecidae). Life-history: Noyes (1922).

38. *Proales werneckii* (Ehrenberg, 1834)

Figs. 287-295

Ehrenberg 1834: 216 (*Notommata werneckii*); Ehrenberg 1838: (*Copeus werneckii*); Hudson & Gosse 1886: 23, pl. 32 fig. 18 (*Proales werneckii*); Rousselet 1895: 415-418, pl. 19 figs. 1-4 (male); Budde 1924: 707-723, figs. 1-6; Harring & Myers 1924: 426-427, pl. 17 figs. 1-5; Kutikova 1970: 490, figs. 704a-e; Koste 1978: 282-283, fig. 27, pl. 86 figs. 13a-i,k; Koste & Shiel 1990: 137-138, figs. 3.4a-e.

Type locality

Near Dassau, Germany.

Description

Juvenile female elongate, slender, later fusiform; when mature the body swells enormously by the accumulation of undigested food and amictic eggs, and becomes almost spherical or shapeless; cuticle soft, flexible, outline changing; very transparent. Head offset by shallow neck-fold, longer than wide, with semi-circular rostrum-like prominence. Foot short, c. 1/9 total length, ± slender, 2 pseudosegments. Toes moderately long, ± slender, acute, slightly decurved ventrally. Corona very oblique, laterally 2 strongly ciliated areas, ciliation of buccal field not extending beyond mouth, circumapical band absent. Brain large, saccate. Retrocerebral sac present. No constriction between stomach and intestine. Gastric glands large, spherical. Mastax with very large salivary glands. Eyespot on top or posterior to brain; crystalline body present.

Rami triangular, decurved posteriorly, inner margins set apart. Fulcrum ± rod-shaped, slightly expanding posteriorly. Unci with single tooth, expanded into a triangular basal plate. Manubria long, slender, posterior end hook-shaped, head with small posterior and anterior lamella. Epipharynx 2 sigmoid plates with rib, near base of rami.

Resting egg (fig. 293) elongate oval, smooth.

Length 140-200 μm, toe 11-16 μm; trophi 12-18 μm; subitaneous egg 56-78x42-52 μm, resting egg 62-72x50-61 μm.

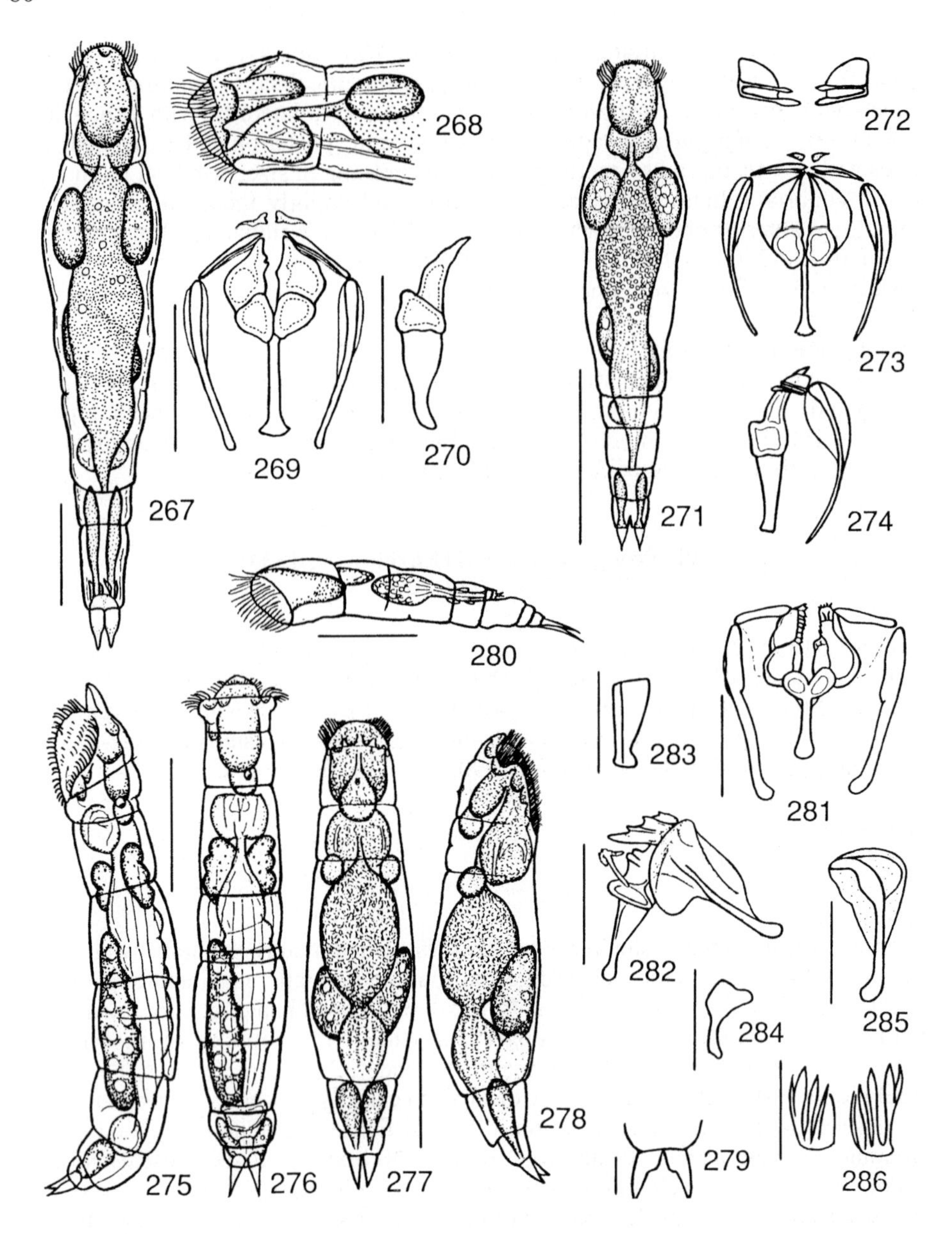

Figs. 267-270: *Proales adenodis*. 267: dorsal view, 268: head, lateral view, 269: trophi, ventral view, 270: incus, lateral view.
Figs. 271-274: *Proales granulosa*. 271: dorsal view, 272: unci, 273: trophi, ventral view, 274: trophi, lateral view.
Figs. 275-286: *Proales decipiens*. 275: lateral view, 276: dorsal view, 277: dorsal view, 278: lateral view, 279: toes, 280: male, 281: trophi, ventral view, 282: trophi, lateral view, 283: fulcrum, lateral view, 284: epipharyngeal element, 285: manubrium, 286: unci.
Scale bars: habitus 50 μm (279: 10 μm), trophi 10 μm.
(267-274: after Myers, 1933a; 275,276,280,281,283-286: after Wulfert, 1961; 279: after Wulfert, 1968; 287: after Harring & Myers, 1922)

Male (fig. 294) with functional mastax and rudimentary digestive system. Length 128-150 μm.

Distribution and ecology
Europe, N. America, Australia. Parasitic in filaments of aquatic *Vaucheria* spp. and *Dichotomosiphon tuberosus* (Braun) Ernst (Xanthophyta); also in *Vaucheria terrestris* (Vauch.) De Candolle growing on moist soil. The rotifer usually enters the developing gametophore where it induces gall formation; here it feeds on the cytoplasm and the cytoplasmic organelles. Egg deposition is within the galls. Young animals leave the gall; copulation is outside the host, with resting eggs produced overwintering in the sediments. Literature: Budde (1924), Davis & Gworek (1973), Christensen (1987), Del Grosso (1980, 1988), May (1989).

Note
It is unknown if gall formation is only caused by *P. werneckii*. On several occasions (e.g. Debray, 1890; Rieth, 1980) resting eggs ornamented with spines, instead of the smooth ones of *P. werneckii*, have been observed in *Vaucheria*. This could point towards another species (Voigt, 1957) or the existence of pseudosexual eggs.

39. *Proales parasita* (Ehrenberg, 1838)
Figs. 296-306

Ehrenberg 1838: 426, pl. 50 fig. 1 (*Notommata parasita*); Rousselet 1911: 8 (*Proales parasita)*; Harring & Myers 1922: 607-608, pl. 51 figs. 1-4; Wesenberg-Lund 1923: ? male; Budde 1924: 723-726, fig. 7; Wulfert 1960: 326, figs. 52a-d; Kutikova 1970: 499, figs. 709a-d; Koste 1978: 283, pl. 86 figs. 5a-f, pl. 92 figs. 8a-d; Koste & Shiel 1990: 136, figs. 3.1a-d.

Type locality
Near Berlin, Germany.

Description
Body of free-swimming individuals fusiform or roughly cylindrical, tapering towards foot; parasitizing specimens more shapeless; cuticle soft, outline fairly constant. Head and neck offset by transverse folds. Trunk with pseudosegments. Tail a rounded median lobe. Foot short, c. 1/12-1/18 total length, broad, with 2 pseudosegments. Toes short, conical, ± abruptly tapering to acute, perforated points. Corona slightly oblique, 2 lateral, strongly ciliated areas. Brain large, saccate; eyespot at distal end: red pigment granules in small vesicle (= rudimentary retrocerebral sac?), sometimes displaced to right. Stomach separated from intestine, filled with green or dark red-greenish material. Gastric glands small, globular. Pedal glands large, sausage-shaped, with reservoirs.

Trophi modified virgate, small. Rami triangular, slightly asymmetrical, right ramus somewhat more developed; teeth on inner margins; alulae absent; basal apophyses nearly semi-circular plates. Fulcrum long, straight, slender, in ventral view rod-shaped; laterally gradually tapering to posterior end. Unci 3-toothed. Manubria ± long, proximal half with posterior and anterior lamella. Epipharynx 2 thin, slightly bent rods.

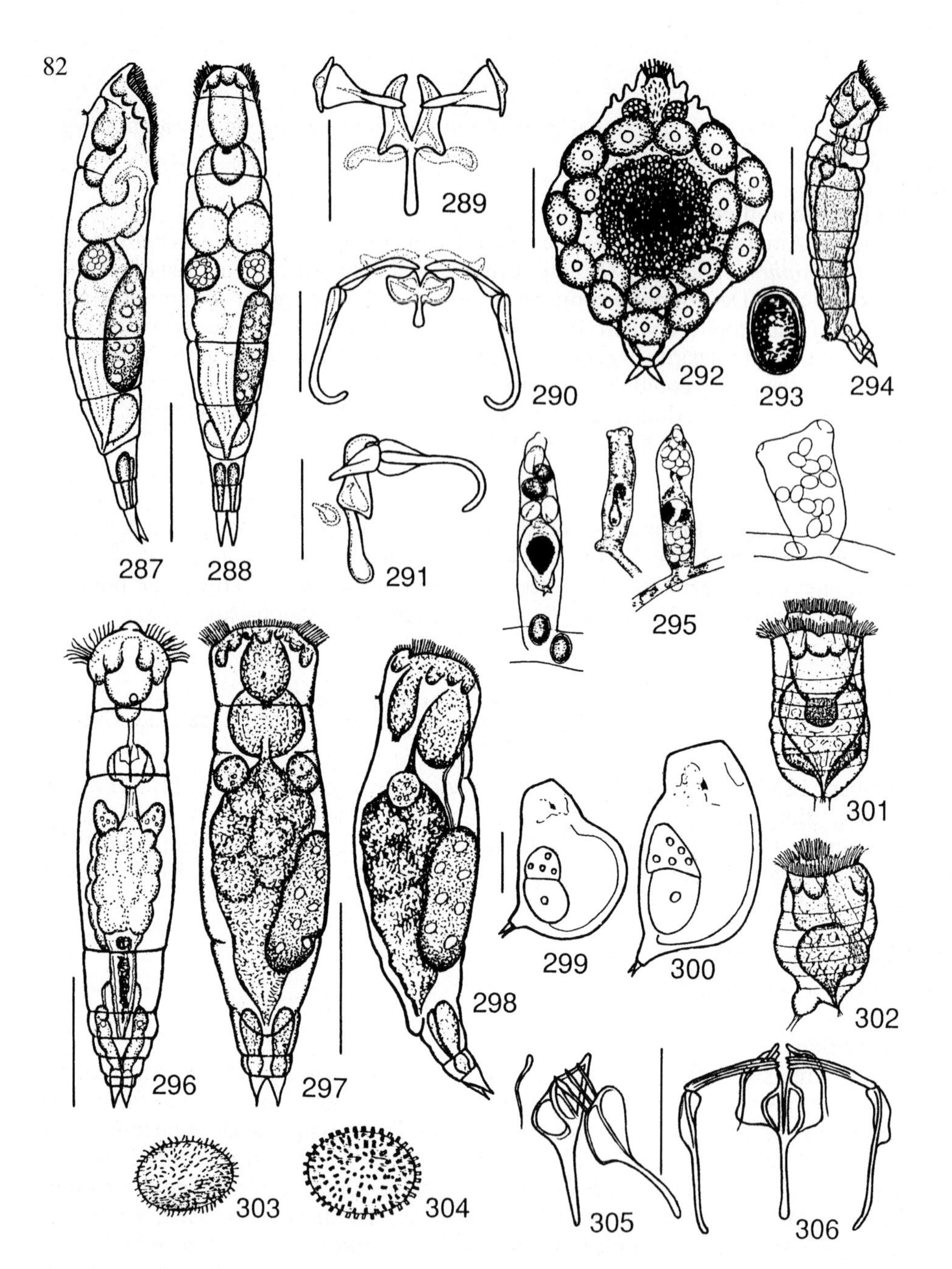

Figs. 287-295: *Proales werneckii*. 287: free-swimming female, lateral view, 288: free-swimming female, dorsal view, 289: trophi, apical view, 290: trophi, ventral view, 291: trophi, lateral view, 292: parasitic, mature female, 293: resting egg, 294: male, lateral view, 295: galls in *Vaucheria*.
Figs. 296-306: *Proales parasita*. 296: free-swimming female, dorsal view, 297: free-swimming female, dorsal view, 298: free-swimming female, lateral view, 299-300: outline of parasitic, mature females, 301: ? male, dorsal view, 302: ? male, lateral view, 303,304: resting eggs, 305: trophi, lateral view, 306: trophi, ventral view.
Scale bars: habitus 50 μm, trophi 10 μm. (287-291: after Harring & Myers, 1924; 292,293: after Balbiani, 1878; 294: after Rousselet, 1897, 295: after Rieth, 1980; 296: after Wulfert, 1960; 297,298, 305,306: after Harring & Myers, 1922; 299-302: after Wesenberg-Lund, 1923; 303,304: after Budde, 1924)

Two kinds of resting eggs (fertilized and pseudosexual eggs?) have been reported (Budde,1924): brown eggs (fig. 303) with long, curved spines (common) and eggs (fig. 304) with short, thick and less numerous spines (scarce).

Length 140-180 µm, toe 10 µm; trophi 15 µm; subitaneous egg 60 µm, male egg 52 µm, long-spined resting egg 65x50 µm (spine 5 µm).

The males (figs. 301,302) described by Wesenberg-Lund (1923) and attributed to the species, probably do not belong to *P. parasita* (Voigt, 1957).

Distribution and ecology

Europe, Asia, N.America, Australia. Parasitic in colonial algae (*Volvox, Uroglena, Uroglenopsis*) and on colonies of ciliates (*Ophrydium*). Feeding behaviour, parasitism: Budde (1924), Carlier (1935), May (1989), Sauer (1978).

40. *Proales lenta* Wlastow, 1956

Figs. 307-309

Wlastow 1956: 668-672, figs. 1-5; Kutikova 1970: 492, figs. 708a,b.

Type locality

Bolshevsk, in pool fed by ground water, near Biological Station of the State University of Moscow, Russia.

Description

Body fusiform, stout. Head offset by distinct neck-fold, ± quadratic, no rostrum. Trunk moderately arched dorsally, posterior margin with broad, truncate tail. Foot relatively small, c. 1/6 total length, 3 indistinct pseudosegments. Toes short, spread, ± cylindric, with ± abruptly reduced, blunt tips. Corona slightly oblique, no lateral auricle-like tufts. Brain saccate. Retrocerebral sac posterior to brain. Eyespot absent. Gastric glands rounded. No constriction between stomach and intestine. Pedal glands elongated, reservoirs in toes.

Rami outline ± quadratic, with protruding acute corners, inner margins and basal apophyses with stout teeth. Fulcrum long, in ventral view rod-shaped, slightly expanding posteriorly. Unci asymmetric: right a single, bifurcate principal tooth and 3 rudimentary toothlets, left a single simple tooth and 2 rudimentary toothlets; small subunci; manubria stout, straight.

Length 425 µm, toe 33 µm; ramus 22.5x27.5 µm, fulcrum 32.5 µm, uncus 15 µm, manubrium 32.5 µm.

Distribution and ecology

Russia. In shallow pools; October-November.

Note

The general appearance and trophi of *P. lenta* remember *Resticula gelida* (Harring & Myers, 1922); cf. also with figs. of *R. gelida* in Wulfert (1935: 595-596, fig. 11); synonymous?

41. *Proales sigmoidea* (Skorikov, 1896)
Figs. 310-318

Synonym: *P. macropoda* Zavadowski, 1926

Skorikov 1896: 284, pl. 7 fig. 8 (*Pleurotrocha sigmoidea*); Zavadowski 1926: 270, fig. 20 (*Pleurotrocha macropoda*); Fadeew 1927: 3, pl. 1 figs. 5-7 (*Proales sigmoidea*); De Beauchamp 1948: 136-138; Kutikova 1962: 479, figs. 7a-e; Koste 1968b: 240-245, figs. 1-7; Koste 1970a: 150, pl. 7 figs. 1-7; Kutikova 1970: 492, figs. 710a-e; Koste 1976: 210, pl. 21 figs. 1a,b; Koste 1978: 276, pl. 90 figs. 1a-h.

Type locality
Near Moscow, Russia.

Description
Body fusiform, short, stout. Head short, distinctly offset by neck-fold. Trunk somewhat inflated, dorsally arched, a few dorsal transverse folds. Tail short, broad. Foot medium long, c. 1/5 total length, stout, 3 pseudosegments. Toes short, conical in lateral view, ± foliate dorsally, ± acutely pointed. Corona oblique. Eyespot red, at posterior of brain. No constriction between stomach and intestine; intestine short. Gastric glands large, rounded. Pedal glands large, pyriform; reservoirs with slender ducts. Salivary glands unicellular.

Trophi virgate. Rami triangular, with long, dorsally curved projections anteriorly; inner margins of projections basally denticulate, anteriorly with double rows, delicate denticles; alulae distinct; basal apophyses small, single-toothed. Fulcrum in ventral view rod-shaped, slender, posterior end expanded and scalloped; laterally broader. Unci curved plates with 2 distinct principal teeth and 7-8 ribs. Manubria long, rod-shaped, slightly curved, with small ventral prominence, posterior and anterior lamellae small.

Length 330-550 μm, toe 20-32 μm; trophi 42 μm: fulcrum 19-20 μm.

Distribution and ecology
Europe (Germany, France, Russia), N. America (Canada). In colonies of sessile ciliates (e.g. *Vorticella*), stagnant and running waters; 5.9-8.7 °C, pH 6.6-7.7. Food: *Campanella umbellaria* L.

42. *Proales daphnicola* Thompson, 1892
Figs. 330-336, pl. 13 figs. 1-5

Synonym: *P. nova* Wlastow, 1953
P. pejleri De Smet, Van Rompu & Beyens, 1992 (n. syn.)

Thompson 1892: 220, fig. 125; Harring 1913: 84 (*Pleurotrocha daphnicola*); Myers 1917: 478, pl. 41 figs. 4-9 (*Pleurotrocha daphnicola*); Harring & Myers 1924: 430-431, pl. 18 figs. 1-5 (*Proales daphnicola*); Wulfert 1939: 576-577, figs. 8a-d (*Pleurotrocha daphnicola*); Hollowday 1949a: 1-7; Wlastow 1953a: 1110-1112, figs. 1-3 (*P. nova*); Wlastow 1954: 52-64, figs. 1-8; Wlastow 1955: 80-84, figs. 1-5 (male); Wulfert 1959: 63, figs. 15a-d (*Pleurotrocha daphnicola*); Koste 1968b: 243-244, figs. 8-11; Koste 1970b: 49-51, figs. 1-7; Kutikova 1970: 494, figs. 711a-g; Koste 1976: 210, pl. 21 figs. 3a,b; Koste 1978: 275-276, figs. 26.3, pl. 86 figs. 8a-i, pl. 90 figs. 2a-g, 5a-c; Koste & Shiel 1990: 133, figs. 1.5a-d .

Type locality
England.

Description
Body broadly fusiform, stout; cuticle soft, flexible, outline fairly constant; moderately hyaline, colourless. Head short, offset by 1-2 distinct neck-folds. Trunk somewhat inflated, arched dorsally; sometimes with dorsal transverse folds. Tail broad, less prominent. Foot moderately long, c. 1/4-1/5 total length, stout, 2 pseudosegments (often indistinct). Toes short, stout, bluntly conical with tubular points. Corona slightly oblique, marginal ciliation ± weak, laterally 2 areas with long cilia, dorsal margin of buccal field with tentacles. No constriction between stomach and intestine. Gastric glands large, rounded, laterally compressed. Pedal glands large, pyriform, laterally compressed, with reservoirs. Brain very large, saccate. Retrocerebral organ absent. Eyespot small, red, occasionally colourless in old specimens, at underside of brain.

Trophi malleate, robust. Rami symmetrical, broad, trapezoid to triangular in dorsal view, strongly curved in lateral view; lateral margins tapering to fulcrum, terminating in acute projections pointing to posterior; anterior margins oblique, each with short, curled projection on antero-median corner; a ± large, rounded dorsal fenestra near anterior margins; ventral surface densely set with skleropili. Basal apophyses rami length, horn-shaped, inner margins finely serrate; each with large, rounded postero-lateral openings at their base. Fulcrum medium long, in ventral view slightly expanding towards posterior end; laterally very broad with slightly expanded and scalloped posterior end. Unci with (5)-6 acute teeth, gradually decreasing in size, first 3 teeth slightly clubbed, others ± linear, inner side with plate covered by small skleropili and brush of long, acute skleropili at anterior margins. Manubria broad, club-shaped with short, straight cauda; anterior cavity long, posterior cavity, curved downwardly.

Resting egg (fig. 336) oval, ornamented with large, flattened granules.

Length 275-500 μm, toe 25-35 μm; trophi 36-40 μm, uncus up to 18 μm; subitaneous egg 96x30 μm, resting egg 105-109x76-80 μm.

Male (fig. 332) similar in habitus as female. Length 140-180 μm.

Distribution and ecology
Wide-spread (Europe, Asia, N. America, Africa, Australia). Epizoic on *Daphnia* spp., rarely on other cladocerans, copepods, amphipods and oligochaetes (?); feeding on unicellular euglenaceans (e.g. *Colacium vesiculosum* Ehrbg.) and ciliates living on carapace. Eggs generally are attached between the basal segments of the 2nd antennae. Literature: Hollowday (1949a), Matveeva (1989), May (1989), Pourriot (1965), Wlastow (1953b).

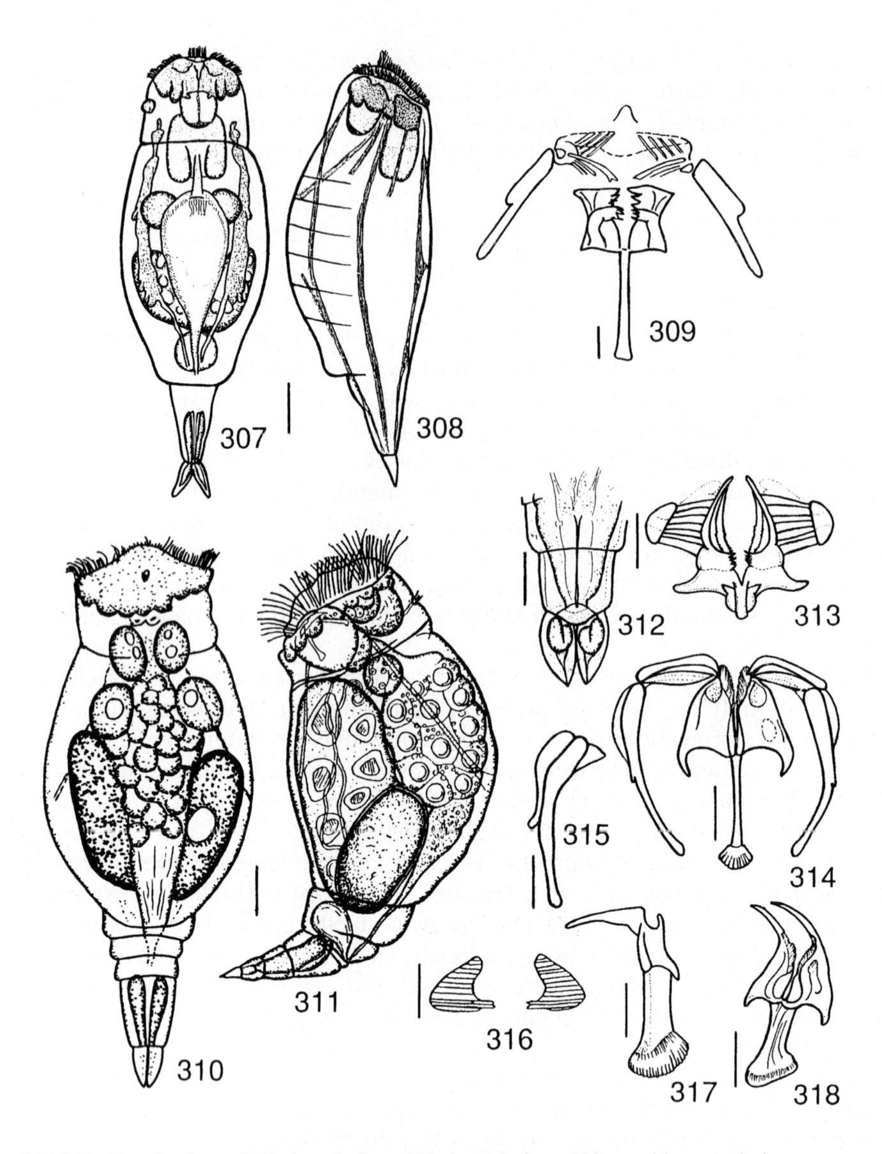

Figs. 307-309: *Proales lenta*. 307: dorsal view, 308: lateral view, 309: trophi, ventral view.
Figs. 310-318: *Proales sigmoidea*. 310: dorsal view, 311: lateral view, 312: toes, 313: trophi, apical view, 314: trophi, ventral view, 315: manubrium, 316: unci, 317: incus, lateral view, 318: incus, lateral oblique view.
Scale bars: habitus 50 μm (312: 20 μm), trophi 10 μm.
(307-309: after Wlastow, 1956; 310: after Kutikova, 1962; 311,315: after Fadeew, 1927; 312-314,317,318: after Koste 1968b, 316: after Kutikova, 1970)

43. *Proales kostei* Nogrady & Smol, 1989
Figs. 319-329, pl. 14 figs. 1-5

Nogrady & Smol 1989: 239, figs. 6a-g.

Type locality and type
Horseshoe and Beach Ridge ponds, Cape Herschel, Ellesmere Island, Northwest Territories, Canada. Holotype in the National Museum of Natural Sciences, Ottawa, Ont.

Description
Body fusiform, stout, rounded in cross-section, semicontracted animals bent; cuticle fairly thick. Head distinctly offset by neck-fold, slightly tilted ventrally. Trunk with pseudosegments and deep furrows, continuing directly into foot. Foot short, with 2 pseudosegments. Toes short, conical, bluntly pointed. Corona frontal, oblique, convex, uniformly ciliated. Intestine indistinct. Gastric glands pyriform, stalked. Pedal glands small, club-shaped.

Trophi malleate. Rami broad, outer margins ± parallel-sided 3/5 their length, tapering anteriorly and posteriorly; rami with pronounced, rounded corners at 1/5 from fulcrum, dorsally terminating in blunt projections pointing to posterior; antero-dorsal tips with stout skleropili; median rami opening pyriform, in anterior half; each ramus with elongate dorsal fenestra. Basal apophyses broad, hollow, spoon-shaped, each with large, elongate opening at their base; inner margins of spoon-shaped part finely denticulate. Fulcrum relatively short, in ventral view slightly expanding towards posterior end; laterally very broad with slightly expanded and scalloped posterior end. Unci with 5-6 teeth, first 3 slightly clubbed, last 2 linear or slightly clubbed. Manubria broad, club-shaped, anterior cavity long, posterior cavity shorter, curved downwardly.

Length (bent animal) 280-320 μm, width 130-160 μm, toe 10-30 μm; trophi 33-43 μm: fulcrum 10-12 μm, unci 20 μm, manubrium 24-25 μm.

Distribution and ecology
Canadian Arctic (Devon Island, Ellesmere Island, Victoria Island). Among detritus and submerged mosses in ponds. July. pH 7.3-8.2, 199-224 μSm^{-1}.

Species inquirendae, nomina dubia and reallocations

Proales algicola: Kellicott, 1897: 48. ? *Cephalodella catellina* (O.F. Müller) according to Harring en Myers (1924). Insufficiently described; nomen dubium.

Proales aureus: Zavadowski, 1916: 278, pl. 4 figs. 1-9. Parasitic in colonies of *Volvox aureus*. Insufficiently described; nomen dubium.

Proales coryneger: Gosse, 1887b: 863, pl. 14 fig. 4. Insufficiently described; nomen dubium.

Proales gammari: Plate, 1886: 236, pl. 7 fig. 42. Synonymous with *Proales*

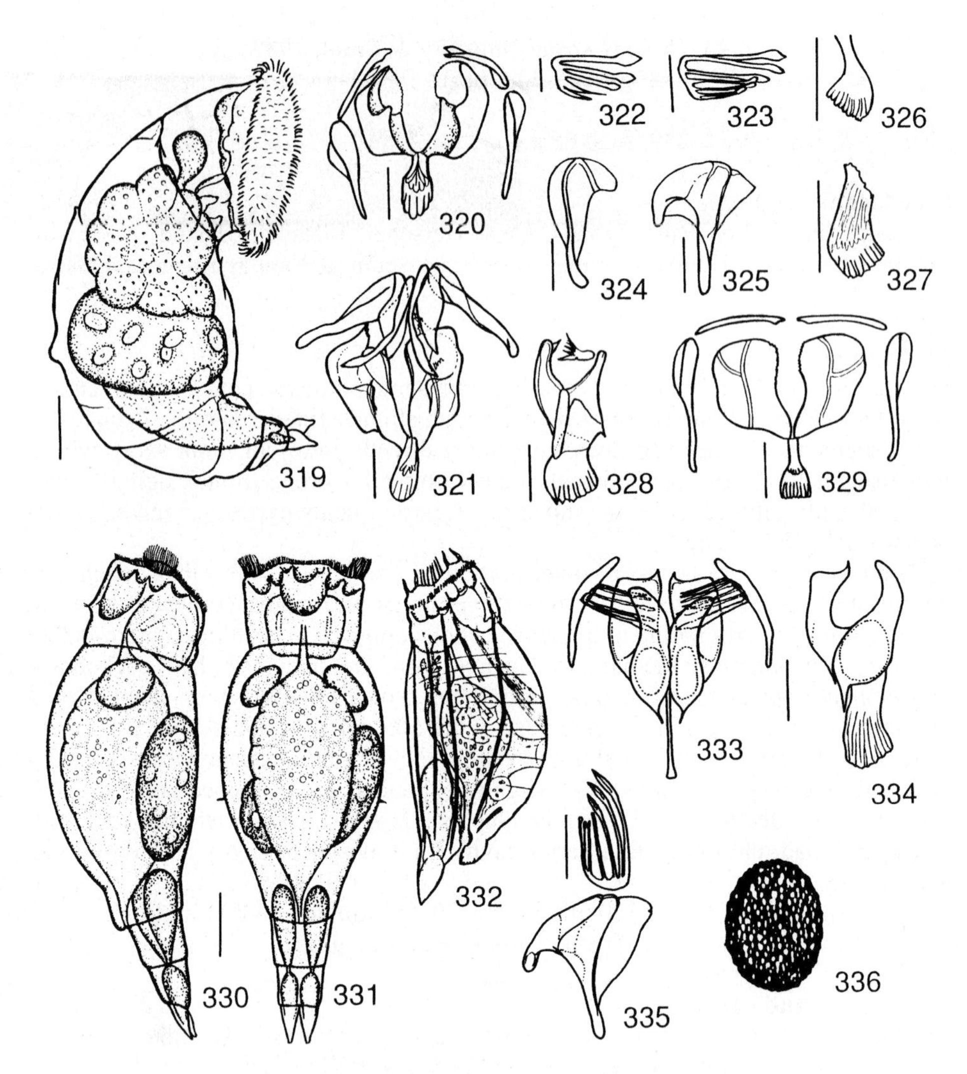

Figs. 319-329: *Proales kostei*. 319: lateral view, 320,321: trophi, dorsal view, 322,323: unci, 324,325: manubrium, 326,327: fulcrum, lateral view, 328: incus, lateral view, 329: trophi, ventral view.
Figs. 330-336: *Proales daphnicola*. 330: lateral view, 331: dorsal view, 332: male, 333: trophi, ventral view, 334: trophi, lateral view, 335: malleus, 336: resting egg.
Scale bars: habitus 50 μm, trophi 10 μm. (319,320,322,324,326,332: after Nogrady & Smol, 1989; 321,323,325,327,328: Devon Island, N.W.T., Canada; 330,331: after Harring & Myers, 1924; 332: after Wlastow, 1955; 333-335: Boom, Belgium; 336: after Voigt, 1957)

reinhardti (Ehrenberg) according to Harring & Myers (1924) and Koste (1978), with *P. daphnicola* (Thompson) according to De Beauchamp (1923), or with *P. theodora* after Hauer (1938); nomen dubium.

Proales halophila sensu Tzschaschel, 1979: 12-15, figs. 4a-e, 5a-c. See *P. halophila*; species inquirenda.

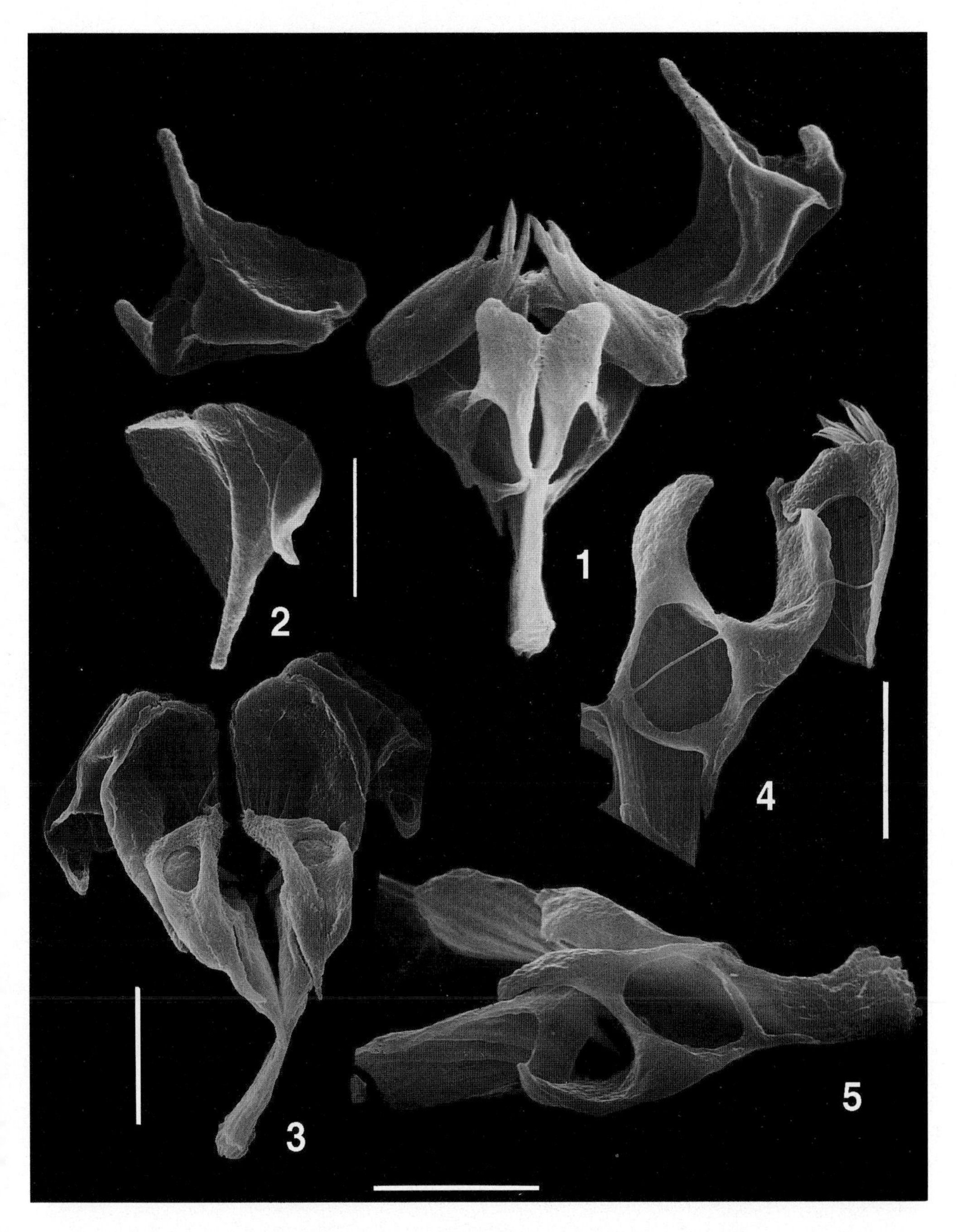

Plate 13. *Proales daphnicola*, trophi (S.E.M. photographs). 1: ventro-apical view, 2: manubrium, 3: dorso-caudal view, 4: detail ramus and inner side of uncus, 5: lateral view, detail incus.
Scale bars: 10μm.
(1,3,5: Boom, Antwerpen, Belgium; 2,4: Blokkersdijk, Antwerpen, Belgium)

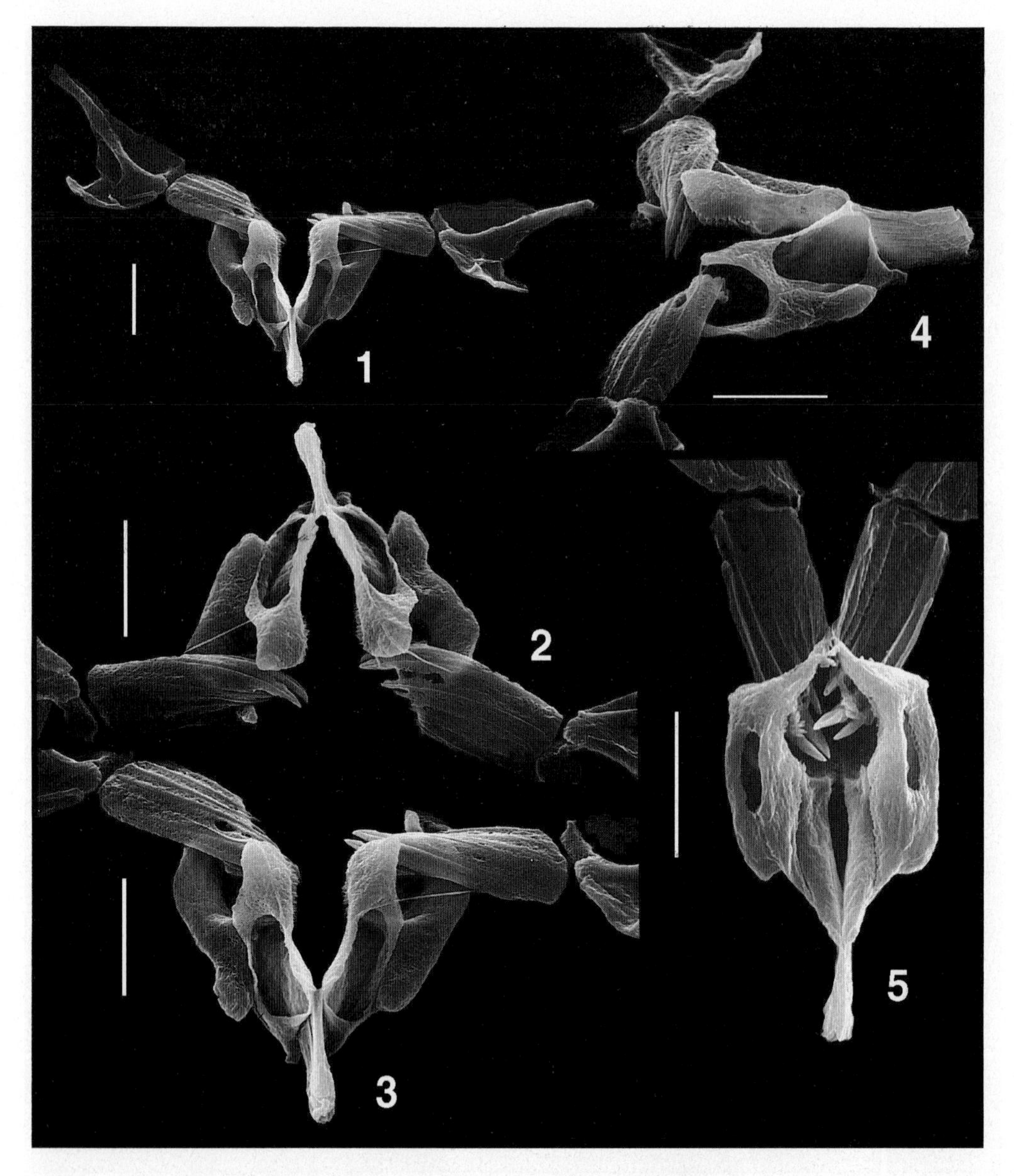

Plate 14. *Proales kostei*, trophi (S.E.M. photographs). 1: ventro-apical view, 2: detail incus and unci, apical view, 3: idem, ventral view, 4: idem, lateral view, 5: idem, dorsal view.
Scale bars : 10 μm
(1-5 : Cambridge Bay, Victoria Island, Canada, N.W.T.)

Proales inflata: Glascott, 1893: 51, pl. 4 fig. 1. Insufficiently described; nomen dubium.

Proales latrunculus: Penard, 1909: 142, figs. 1-7. In *Acanthocystis turfacea* Carter (Heliozoa). Insufficiently described; species inquirenda.

Proales longidactyla: Edmondson, 1948: 150, figs. 8-13. According to Segers (1993) synonymous with *Lecane clara* (Bryce).

Proales namibiensis: Koste & Brain, 1993 *in* Brain & Koste 1993: 451-454, figs. 4a,b, 5c-h. Synonymous with *Lecane hastata* (Murray) according to Segers (1995).

Proales orthodon: Gosse, 1887a: 366, pl. 8 fig. 11. Insufficiently described; nomen dubium.

Proales prehensor: Gosse, 1887a: 366, pl. 8 pl. 12. Synonymous to *Lecane levistyla* (Olofsson) according to Harring & Myers (1924).

Proales psammophila: Neiswestnowa-Shadina, 1935: 560-561, fig. 5,6 = *Proales minima* f. *psammophila* Koste (1978); synonymous to *P. minima* according Wiszniewski (1935). Since the trophi are poorly described, it is impossible to make a distinction with *P. minima* (Montet); nomen dubium.

Proales tyrphosa: Bērziņš, 1948: 315-317, figs. 6-11. Description insufficient, trophi poorly illustrated. According to Kutikova (1970) probably synonymous with *P. fallaciosa* Wulfert.

Proales uroglenae: De Beauchamp, 1948: 138-140, figs. 1A-C. Examination of specimens revealed that the taxon does not belong to *Proales*. The external morphology, internal organization and the characteristic slender trophi, with lyrate rami and curved manubria with ventrally projecting lamella, suggest that the species is closely related to *Pleurotrocha chalicodis* Myers, 1933, *P. trypeta* (Harring & Myers, 1922) and *P. vernalis* Wulfert, 1935 (Notommatidae). *Proales uroglenae* is therefore reallocated to *Pleurotrocha uroglenae*, depending the revision of the notommatid genera.

CHECK-LIST OF NAMES AND SYNONYMS

Names for valid species are printed in bold; synonyms, species inquirendae and nomina dubia indented.

Bryceella

  B. agilis Neiswestnowa-Shadina, 1935: sp. inq.

B. stylata (Milne, 1886)

B. tenella (Bryce, 1897)

  B. voigti Rodewald, 1935: sp. inq.

Proales

P. adenodis Myers, 1933

P. alba Wulfert, 1939

  P. algicola Kellicott, 1897: nom. dub.

  P. aureus Zavadowski, 1916: nom. dub.

P. baradlana Varga, 1958

P. bemata Myers, 1933

  P. brevipes Harring & Myers, 1924: see *P. decipiens*

P. christinae De Smet, 1994

P. cognita Myers, 1940

P. commutata Althaus, 1957

  P. coryneger Gosse, 1887: nom. dub.

P. cryptopus Wulfert, 1935

P. daphnicola Thompson, 1892

P. decipiens (Ehrenberg, 1832)

P. doliaris (Rousselet, 1895)

P. fallaciosa Wulfert, 1937

  P. gammari Plate, 1886: nom. dub.

P. germanica Tzschaschel, 1978

P. gigantea (Glascott, 1893)

P. gladia Myers, 1933

P. globulifera (Hauer, 1921)

  P. globulifera var. *halophila* Remane, 1929: see *P. halophila*

P. gonothyraeae Remane, 1929

P. granulosa Myers, 1933

P. halophila (Remane, 1929)

  P. halophila sensu Tzschaschel, 1979: sp. inq.

P. indirae Wulfert, 1966

  P. inflata Glascott, 1893: nom. dub.

P. kostei Nogrady & Smol, 1989

  P. latrunculus Penard, 1909: sp. inq.

  P. longidactyla Edmondson, 1948 = *Lecana clara* (Bryce, 1892)

P. lenta Wlastow, 1956

  P. longipes Remane, 1929: see *P. theodora*

  P. macropoda Zavadowski, 1926: see *P. sigmoidea*

P. macrura Myers, 1933

P. micropus (Gosse, 1886)

P. minima (Montet, 1915)
P. minima f. *psammophila* Koste, 1978: nom. dub.
P. namibiensis Koste & Brain, 1993 = *Lecane hastata* (Murray, 1913)
P. nova Wlastow, 1953: see *P. daphnicola*
P. oculata Tzschaschel, 1978
P. ornata Myers, 1933
P. orthodon Gosse, 1887: nom. dub.
P. ovicola Giard, 1908: see *P. gigantea*
P. paguri Thane-Fenchel, 1966
P. palimmeka Myers, 1940
P. parasita (Ehrenberg, 1838)
P. pejleri De Smet, Van Rompu & Beyens: see *P. daphnicola*
P. phaeopis Myers, 1933
P. prehensor Gosse, 1887 = *Lecane levistyla* (Olofsson, 1917)
P. provida Wulfert, 1938
P. psammophila Neiswestnowa-Shadina, 1935: nom. dub.
P. pugio Nogrady, 1983
P. quadrangularis Glascott, 1893: see *P. globulifera*
P. reinhardti (Ehrenberg, 1834)
P. segnis Myers, 1938
P. sigmoidea (Skorikov, 1896)
P. similis De Beauchamp, 1907
P. similis var. *exoculis* Bērziņš, 1953: syn.
P. simplex Wang, 1961
P. sordida Gosse, 1886
P. syltensis Tzschaschel, 1979
P. theodora (Gosse, 1887)
P. theodora var. *calcarata* Wulfert, 1938: syn.
P. tyrphosa Bērziņš, 1948: see *P. fallaciosa*
P. uroglenae De Beauchamp, 1948 = *Pleurotrocha uroglenae* n. comb.
P. werneckii (Ehrenberg, 1834)
P. wesenbergi Wulfert, 1960

Proalinopsis
P. caudatus (Collins, 1873)
P. gracilis (Myers, 1933)
P. lobatus Rodewald, 1935: sp. inq.
P. montanus Godeanu, 1963: sp. inq.
P. phacus Myers, 1933
P. selene Myers, 1933
P. squamipes Hauer, 1935
P. staurus Harring & Myers, 1924
P. trisegmentus Sudzuki, 1960: sp. inq.

Wulfertia
W. kindensis Koste & Tobias, 1990
W. kindensis kivuensis De Smet, 1992: see *W. kivuensis*
W. kivuensis De Smet n. stat.
W. ornata Donner, 1943

REFERENCES

Althaus, B. 1957. Neue Sandbodenrotatorien aus dem Schwarzen Meer. Wiss. Z. Univ. Halle, Math. Nat. VI/3: 445-458.

Balbiani, E.G. 1878. Observations sur le notommate de Werneck et sur son parasitisme dans les tubes des Vauchéries. Ann. Sci. Nat., Paris, Zool., ser 6, 7(2): 1-40, 4 pl.

Bartoš, E. 1959. Vířníci - Rotatoria. Fauna ČRS 15: 969 pp., Náklad. Českosl. Akad. Věd, Praha.

Bērziņš, B. 1948. Einige neue Notommatidae-Arten (Rotatoria) aus Schweden. Hydrobiologia 1: 312-321.

Bērziņš, B. 1952. Contributions to the knowledge of the marine Rotatoria of Norway. Univ. Bergen Årbok 1951, Naturvit. rek. 6(3): 1-11.

Bērziņš, B. 1953. Zur Kenntnis der Rotatorien aus West-Australien. Lunds univ. Arsskr., N.F., Avd. 2, 49(8): 9-12.

Björklund, B.G. 1972. The rotifer fauna of rock-pools in the Tvärminne Archipelago, Southern Finland. Acta Zool. Fennica 135: 1-30.

Brain, C.K. & Koste, W. 1993. Rotifers of the Genus *Proales* from saline springs in the Namib desert, with the description of a new species. Hydrobiologia 255/256: 449-454.

Braioni, M.G. & Gelmini, D. 1983. Rotiferi Monogononti (Rotatoria: Monogononta). Guide per il riconoscimento delle specie animali delle acque interne Italiane 23: 180 pp.

Bryce, D. 1897. Contributions to the non marine fauna of Spitsbergen. Part. II. Report on the Rotifera. Proc. Zool. Soc. London (1897): 793-799.

Budde, E. 1924. Die parasitischen Rädertiere mit besonderer Berücksichtigung der in der Umgegend von Minden I.W. beobachteten Arten. Z. Morph. Ökol. Tiere, Abt. A, 3: 706-784.

Carlier, E.W. 1935. Hartlebury Common records. The invertebrate fauna of the pools during the years 1932 and 1933. Proc. Birmingham Nat. Hist. Philos. Soc. 16: 131-141.

Chengalath, R. 1985. The Rotifera of the Canadian Arctic sea ice, with description of a new species. Can. J. Zool. 63: 2212-2218.

Christensen, T. 1987. Some collections of *Vaucheria* (Tribophyceae) from south-eastern Australia. Austr. J. Bot. 35: 617-630.

Collins, F. 1873. New species of Rotatoria. Science-Gossip: 9-11.

Czapik, A. 1952. Untersuchungen über die Infusorien und Rotatorien des Küstengrundwassers und des Sandbodens der Stalin-Bucht. Arb. Biol. Meeresstat. Stalin (Bulgarien) 17: 61-65.

Dartnall, H.J.G. & Hollowday, E.D., 1985. Antarctic rotifers. Brit. Antarctic Surv., Sc. Rep. 100: 1-46.

Davis, J.S. & Gworek, W.F. 1973. A rotifer parasitizing *Vaucheria* in a Florida spring. Trans. Amer. Microsc. Soc. 92: 135-140.

De Beauchamp, P. 1907. Description de trois rotifères nouveaux de la faune française. Bull. Soc. Zool. France, 32: 148-157.

De Beauchamp, P. 1923. Courtes notes sur les rotifères. Ann. Biol. Lacustre 12: 221-228.

De Beauchamp, P. 1948. Sur deux espèces de *Proales* (Rotifères). Bull. Soc. Zool. France 73: 136-140.

Debray, F. 1890. Sur *Notommata Werneckii*, Ehrb., parasite des Vauchérieés. Bull. Sci. France et Belgique (1890): 222-242.

Del Grosso, F. 1985. Sulla formazione di galle in *Dichotomosiphon tuberosus* (A.Br.) Ernst (Chlorophyceae, Dichotomosiphonales), prodotte dal rotifero *Proales wernecki*, Giornale Bot. ital. 119, Suppl. 2: 47-48.

Del Grosso, F. 1988. On the formation of galls in *Dichotomosiphon tuberosus* (A. Br.) Ernst (Chlorophyceae, Dichotomosiphonales), produced by rotifer *Proales wernecki* (Ehrenberg) Hudson et Gosse. Nova Hedwigia 46: 157-164.

De Smet, W.H. 1993. Report on rotifers from Barentsøya, Svalbard (78°30'N). Fauna norv. Ser. A 14: 1-26.

De Smet, W.H. 1994. *Proales christinae* (Rotifera, Proalidae): a new species from the littoral of the North Sea. Belg. J. Zool. 124: 21-25.

De Smet, W.H. & Bafort, J.M., 1992. Additions to the rotifer fauna of Lake Kivu (Zaïre) with description of *Wulfertia kindensis kivuensis* subsp. n. and *Ascomorpha dumonti* sp. n. Biol. Jb. Dodonaea 60: 110-117.

De Smet, W.H., Van Rompu, E.A. & Beyens, L. 1992. Contribution to the rotifer fauna of subarctic Greenland (Kangerlussuaq and Ammassalik area). Hydrobiologia 255/256: 463-466.

Donner, J. 1943. Zur Rotatorienfauna Südmährens. Mit Beschreibung der neuen Gattung *Wulfertia*. Zool. Anz. 143: 21-33.

Donner, J. 1952. Rotatoria (Rädertierchen). In: Eichler, W., Die Tierwelt der Gewächshäuser. Akad. Verlagsgesellschaft, Geest & Portig, K.-G., 7-17.
Donner, J. 1952. Bodenrotatoriën im Winter. Mikrokosmos 42: 29-33.
Donner, J. 1955. Zur Rotatorienfauna Südmährens. Öster. Zool. Z. 5: 30-117.
Donner, J. 1964. Die Rotatorien-Synusien submerser Makrophyten der Donau bei Wien und mehrerer Alpenbäche. Arch. Hydrobiol., Suppl. 27: 227-324.
Donner, J. 1978. Material zur saprobiologischen Beurteilung mehrerer Gewässer des Donau-Systems bei Wallsee und in der Lobau, Österreich, mit besonderer Berücksichtigung der litoralen Rotatorien. Arch. Hydrobiol. Suppl. 52: 117-228.
Dorazio, R.M. 1984. The contribution of longevity to population death rates. Hydrobiologia 108: 239-243.
Dujardin, F. 1841. Histoire naturelle des zoophytes infusoires, comprenant la physiologie et la classification de ces animaux, et la manière de les étudier à l'aide du microscope. Librairie Encyclopédique de Roret, Paris. XII + 684 pp., atlas 14 pp., 22 pl.
Edmondson, W.T. 1948. Two new species of Rotatoria from sand beaches, with a note on *Collotheca wiszniewskii*. Trans. Am. Microsc. Soc. 67: 149-152.
Ehrenberg, C.G. 1832. Ueber die Entwickelung und Lebensdauer der Infusionsthiere, nebst ferneren Beiträgen zu einer Vergleichung ihrer organischen Systeme. Abh. Akad. Wiss., Berlin: 1-154, pl. 1-4.
Ehrenberg, C.G. 1834. Dritter Beitrag zur Erkenntniss grosser Organisation in der Richtung des kleinsten Raumes. Abh. Akad. Wiss., Berlin: 145-336, pl. 1-11.
Ehrenberg, C.G. 1838. Die Infusionsthierchen als vollkommene Organismen. Ein Blick in das tiefere organische Leben der Natur. Leipzig, XVIII + 547 pp., 64 pl.
Fadeew, N.N. 1927. Matériaux pour la connaissance de la faune des rotifères del'U.R.S.S. Trav. Soc. Nat. Kharkov 50: 141-155 (in Russian).
Giard, A. 1908. Un nouveau rotifère (*Proales ovicola*), parasite des pontes des mollusques d'eau douce. Feuilles Jeunes Natural., Paris 38: 184.
Glascott, L.S. 1893. A list of some of the Rotifera of Ireland. Proc. Roy. Dublin Soc., n.ser. 8: 29-86, pl. 3-7.
Godeanu, S. 1963. Neue und bemerkenswerte Rädertiere aus dem Bucegi-Gebirge (Südkarpathen) Rumäniens. Zool. Anz. 170: 374-380.
Godske Eriksen, B. 1969. Rotifers from two tarns in southern Finland, with a description of a new species, and a list of rotifers previously found in Finland. Acta Zool. Fennica 125: 3-36.
Gosse, P.H. 1887a. Twelve new species of Rotifera. J. Roy. Microsc. Soc.: 361-367, pl. 8.
Gosse, P.H. 1887b. Twenty-four more new species of Rotifera. J. Roy. Microsc. Soc.: 861-871, pl. 14-15.
Harring, H.K. 1913. Synopsis of the Rotatoria. Bull. U.S. Nat. Mus., Washington 81: 7-226.
Harring, H.K. & Myers, F.J. 1922. The rotifer fauna of Wisconsin. Trans. Wisconsin Acad. Sci., Arts and Letters 20: 553-662, pl. 41-61.
Harring, H.K. & Myers, F.J. 1924. The rotifer fauna of Wisconsin II. A revision of the notommatid rotifers, exclusive of the Dicranophorinae. Trans. Wisconsin Acad. Sci., Arts and Letters 21: 415-549, pl. 16-43.
Hauer, J. 1921. Rädertiere aus dem Gebiet der oberen Donau. Ein Beitrag zur Kenntnis unserer heimischen Rotatorienfauna. Mitt. Bad. Landesver. Naturkunde u. Naturschutz, Freiburg i. Br., N.F. 1: 177-186.
Hauer, J. 1925. Rotatorien aus den Salzgewässern von Oldesloe (Holstein). Mitt. Geogr. Ges. naturhist. Mus. Lübeck II(30): 152-195.
Hauer, J. 1935. Rotatorien aus dem Schluchseemoor und seiner Umgebung. Ein Beitrag zur Kenntnis der Rotatorienfauna der Schwarzwaldhochmoore. Verhandl. Naturwiss. Ver. Karlsruhe 30: 47-130.
Hauer, J. 1938. Zur Rotatorienfauna Deutschlands (VII). Zool. Anz. 123: 213-219.
Hollowday, E.D. 1949a. Introduction to the study of the Rotifera - IX *Proales daphnicola* Thompson: with reference to commensal and parasitic habits. The Microscope 6: 1-7.
Hollowday, E.D. 1949b. A preliminary report on the Plymouth marine and brackish-water Rotifera. J. Mar. Biol. Ass. U.K. 28: 239-253.
Hudson, C.T. & Gosse, P.H. 1886. The Rotifera; or wheel-animalcules, both British and foreign. Vol. I, VI + 128 pp, 15 pl., Vol. II, 144 pp., pl. 16-30. Longmans, Green, and Co., London.
Jansson, A.-M. 1967. The food-web of the *Cladophora*-belt fauna. Helgol. wiss. Meeresunters. 15: 574-588.

Jennings, H.S. 1901. Synopses of North-American Invertebrates. XVII. The Rotatoria. Am. Nat., Boston 35: 725-777, 9 pl.
Jennings, H.S. & Lynch, R.S. 1928. Age, mortality, fertility and individual diversities in the rotifer *Proales sordida* Gosse. I. Effect of age of the parent on characteristics of the offspring. J. Exp. Zool. 50: 345-407.
Jersabek, C.D. & Schabetsberger, R. 1992a. Taxonomisch-ökologische Erhebung der Rotatorien- und Crustaceenfauna stehender Gewässer des Hohen Tauern.Final Rapport "Forschungsinstitut Gastein-Tauernregion", 165 pp.
Jersabek, C.D. & Schabetsberger, R. 1992b. Taxonomisch-ökologische Erhebung der Rotatorien- und Crustaceenfauna stehender Gewässer des Sengengebirges. Final Rapport "Verein Nationalpark Kalkalpen", 115 pp.
Kellicott, D.S. 1897. The Rotifera of Sandusky Bay (second paper). Proc. Am. Soc. Microsc. 19: 43-54.
Koch-Althaus, B. 1963. Systematische und ökologische Studien an Rotatorien des Stechlinsees. Limnologica (Berlin) 1: 375-456.
Koreneva, T.A. 1958. Rotifer parasitizing egg-masses of Tendipedidae. Zool. Zhur. 37: 290-291. (in Russian).
Koste, W. 1965. Die Rotatorien des Naturdenkmals "Engelbergs Moor" in Druchhorn, Kreis Bersenbrück. Veröff. Naturw. Ver. Osnabrück 31: 49-82.
Koste, W. 1968a. Ueber die Rotatorienfauna des Naturschutzgebietes "Achmer Grasmoor" in Achmer, Kreis Bersenbrück. Veröff. Naturwiss. Ver. Osnabrück 32: 107-160.
Koste, W. 1968b. Ueber *Proales sigmoidea* (Skorikow) 1896 (eine für Mitteleuropa neue Rotatorienart) und *Proales daphnicola* (Thompson) 1892. Arch. Hydrobiol. 65: 240-245.
Koste, W. 1970a. Zur Rotatorienfauna Nordwestdeutschlands. Veröff. Naturwiss. Ver. Osnabrück 33: 139-163.
Koste, W. 1970b. Das Rädertier-Porträt. Das Putzer-Rädertier *Proales daphnicola*. Mikrokosmos 59: 49-51.
Koste, W. 1972. Ueber zwei seltene parasitische Rotatorienarten *Drilophaga bucephalus* Vejdovsky und *Proales gigantea* (Glascott). Osnabrücker Naturw. Mitt. 1: 149-158.
Koste, W. 1976. Ueber die Rädertierbestände (Rotatoria) der oberen und mittleren Hase in den Jahren 1966-1969. Osnabrücker Naturw. Mitt. 4: 191-263.
Koste, W. 1978. Rotatoria. Die Rädertiere Mitteleuropas. Ein Bestimmungswerk, begründet von Max Voigt. Ueberordung Monogononta. 2nd. ed. I. Textband, 673 pp., II. Tafelband, 234 Taf., Gebr. Borntraeger, Berlin, Stuttgart.
Koste, W. & Shiel, R.J. 1986. New Rotifera (Aschelminthes) from Tasmania. Trans. R. Soc. S. Austr. 110: 93-109.
Koste, W. & Shiel, R.J. 1990. Rotifera from Australian inland waters. VI. Proalidae, Lindiidae (Rotifera: Monogononta). Trans. R. Soc. S. Aust. 114: 129-143.
Koste, W. & Tobias, W. 1990. Zur Kenntnis der Rädertierfauna des Kinda-Stausees in Zentral-Burma (Aschelminthes: Rotatoria). Osnabrücker naturwiss. Mitt. 16: 83-110.
Kutikova, L.A. 1962. List of Rotatoria of the Luga District of the Leningrad region. Akad. Nauk U.S.S.R., 31: 463-492 (in Russian).
Kutikova, L.A. 1970. Kolovratki fauny SSSR (Rotatoria) [The rotifer fauna of the USSR]. Opred. faune SSSR 104, 744 pp., Akad. Nauk SSSR, Leningrad (in Russian).
Liebers, R. 1937. Beiträge zur Biologie der Rädertiere. Z. Wiss. Zool. 150: 206-261.
Martin, L.V. 1977. Rotifers in the *Sphagnum* pools on Thursley Common. Part 2. Microscopy 33: 236-241.
Matveeva, L.K. 1989. Interrelations of rotifers with predatory and herbivorous Cladocera: a review of Russian works. Hydrobiologia 186/187: 69-73.
May, L. 1989. Epizoic and parasitic rotifers. Hydrobiologia 186/187: 59-67.
Milne, W. 1886. On the defectiveness of the eye-spot as a means of generic distinction in the *Philodinaea*. Proc. Philos. Soc. Glasgow 17: 134-145, pl. 17, 18.
Montet, G. 1915. Contribution à l'étude des rotateurs du bassin du Léman (Région du Haut-Lac). Rev. Suisse Zool. 23: 251-360, pl. 7-13.
Myers, F.J. 1917. Rotatoria of Los Angeles, California, and vicinity, with descriptions of a new species. Proc. U.S. Nat. Mus. 52: 473-478, pl. 40,41.
Myers, F.J. 1933a. The distribution of Rotifera on Mount Desert Island. Part. II. New Notommatidae of the genera *Notommata* and *Proales*. Am. Mus. Nov. 659: 1-25.
Myers, F.J. 1933b. The distribution of Rotifera on Mount Desert Island. III. New Notommatidae of the

genera *Pleurotrocha*, *Lindia*, *Eothina*, *Proalinopsis*, and *Encentrum*. Am. Mus. Nov. 660: 1-18.

Myers, F.J. 1938. New species of Rotifera from the collection of the American Museum of Natural History. Am. Mus. Nov. 1011: 1-17.

Myers, F.J. 1940. New species of Rotatoria from the Pocono Plateau, with note on distribution. Not. Nat. Acad. Nat. Sci. Philadelphia 51: 1-12.

Neiswestnowa-Shadina, K. 1935. Zur Kenntnis des rheophilen Mikrobenthos. Arch. Hydrobiol. 28: 555-582.

Nekrassow, A.D. 1928. Vergleichende Morphologie der Laiche von Süsswasser Gastropoden. Z. Morph. Ökol. Tiere 12: 1-35.

Nogrady, T. 1983. Some new and rare warmwater rotifers. Hydrobiologia 106: 107-114.

Nogrady, T. & Smol, J.P. 1989. Rotifers from five high arctic ponds (Cape Herschel, Ellesmere Island, N.W.T.). Hydrobiologia 173: 231-242.

Nogrady, T., Wallace, R.L. & Snell, T.W. 1993. Rotifera 1: Biology, Ecology and Systematics. Guides to the Identification of the Microinvertebrates of the Continental Waters of the World 4. (H.J. Dumont & T. Nogrady, eds.) SPB Academic Publishing BV., The Hague, The Netherlands, 142 pp.

Noyes, B. 1922. Experimental studies on the life-history of a rotifer reproducing parthenogenetically (*Proales decipiens*). J. Exp. Zool. 35: 225-255.

Pejler, B. 1962. On the taxonomy and ecology of benthic and periphytic Rotatoria. Investigations in northern Swedish Lapland. Zool. Bidrag, Uppsala 33: 327-422.

Penard, E. 1909. Ueber ein bei *Acanthocystis turfacea* parasitisches Rotatorium. Mikrokosmos 2: 135-143.

Plate, L.H. 1886. Beiträge zur Naturgeschichte der Rotatorien. Jenaische Z. Naturwiss. 19, n. ser., 12: 1-120, pl. 1-3.

Pourriot, R. 1965. Recherches sur l'écologie des rotifères. Vie et Milieu, Suppl. 21: 224 pp.

Remane, A. 1929-1933. Rotatoria. In: H.G. Bronn's Klassen und Ordnungen des Tier-Reichs, Bd. 4, Abt. II/1: 1-577.

Remane, A. 1929a. *Proales gonothyraeae* n. sp., ein an Hydroidpolypen parasitierendes Rädertier. Zool. Anz. 80: 289-295.

Remane, A. 1929b. Rotatoria. In: G. Grimpe & E. Wagler, Die Tierwelt der Nord- und Ostsee. L. XVI, T. VII. e: 1-156.

Rieth, A. 1980. Xanthophyceae. 2. Teil. Süsswasseflora von Mitteleuropa. Gustav Fischer Verlag, Stuttgart, New York, 147 pp.

Rodewald, L. 1935. Fauna Rotiferelor din Bucovina. Sistematica, biologia si raspândirea lor geografica. Bul. Fac. Stiinte Cernauti 8: 186-266.

Rousselet, C.F. 1895. On *Diplois trigona* sp. n. and other rotifers. J. Quekett Microsc. Club, ser. 2, 6: 119-126, pl. 6-7.

Rousselet, C.F. 1897. On the male of *Proales wernecki*. J. Quekett Microsc. Club, ser. 2, 6: 415-418, pl. 19.

Rousselet, C.F. 1911. Rotifera (excluding Bdelloida). Proc. Roy. Irish Acad., Dublin 31(51): 10 pp.

Rudescu, L. 1960. Trochelminthes. Rotatoria. Fauna Republ. Pop. Romîne, Vol. 2, Fasc. 2., Acad. Republ. Pop. Romîne, 1192 pp.

Rudescu, L. 1961. Rotiferii din Marea Neagra. Hidrobiologia, Luc. Com. Hidrol. Hidrobiol. Ihtiol. 3: 283-329.

Sauer, F. 1978. Ein Rädertier als Parasit bei *Volvox*. Mikrokosmos 67: 110-111.

Schulte, H. 1959. Beiträge zur Kenntnis der Rädertiefauna des Speicherseegebietes bei München. Zool. Anz. 163: 178-189.

Segers, H. 1993. Rotifera of some lakes in the floodplain of the River Niger (Imo State, Nigeria). I. New species and other taxonomic considerations. Hydrobiologia 250: 39-61.

Segers, H. 1995. Rotifera 2: The Lecanidae (Monogononta). Guides to the Identification of the Microinvertebrates of the Continental Waters of the World 6. (H.J. Dumont & T. Nogrady eds.) SPB Academic Publishing BV., The Hague, The Netherlands, 226pp.

Skorikov, A.S. 1896. Rotateurs des environs de Kharkow. Trav. Soc. Nat. Kharkow 30: 207-374, pl. 7-9 (in Russian).

Stevens, J. 1907. The Rotifera of the Exeter district. Proc. Coll. Field Club and Nat. Hist. Soc., Exeter (1907): 30-52.

Stevens, J. 1912. Note on *Proales* (*Notommata*) *gigantea* Glascott, a rotifer parasitic in the eggs of the Water-snail. J. Quekett. Microsc. Club., ser. 2, 11: 481-486, pl. 24.

Sudzuki, M. 1960. Limnological survey of the Lake Nojiri on the Tableland in Middle Japan. II. Description of some new species of rotifers. Bull. Biogeogr. Soc. Japan 22: 19-26.

Thane-Fenchel, A. 1966. *Proales paguri* sp. nov., a rotifer living on the gills of the hermit crab *Pagurus bernhardus* (L.). Ophelia 3: 93-97.

Thane-Fenchel, A. 1968. Distribution and ecology of non-planktonic brackish-water rotifers from Scandinavian waters. Ophelia 5: 273-297.

Thompson, P.G. 1892. Notes on the parasitic tendency of rotifers of the genus *Proales*; with an account of a new species. Sci. Gossip 28: 219-221.

Tzschaschel, G. 1979. Marine Rotatoria aus dem Interstitial der Nordseeinsel Sylt. Mikrofauna Meeresboden 71: 1-64.

Tzschaschel, G. 1980. Verteilung, Abundanzdynamik und Biologie mariner interstitieller Rotatoria. Mikrofauna Meeresboden 81: 1-56.

Varga, L. 1958. Beiträge zur Kenntnis der aquatilen Mikrofauna der Baradla-Höhle bei Aggtelek. Acta Zool. Acad. Sci. Hung. 4: 429-441.

Voigt, M. 1957. Rotatoria. Die Rädertiere Mitteleuropas. I. Textband, 508 pp., II. Tafelband, 115 Tafeln. Gebrüder Borntraeger, Berlin.

Von Hofsten, N. 1910. Rotatorien aus dem Mästermyr (Gottland) und einigen andern schwedischer Binnengewässern. Ark. Zoologi, Stockholm 6: 1-251.

Von Hofsten, N. 1912. Marine, litorale Rotatorien der skandinavischen Westküste. Zool. Bidrag, Uppsala 1: 163-228.

Wang Chia-Chi. 1961. Freshwater rotifers of China. Inst. Freshwat.-Hydrobiol., ANKNR, Peking, I-XI, 1-228 (in Chinese).

Weber, E.F. & Montet, G. 1918. Rotateurs. Catalogus Invertebrés Suisse, Genève 11, XII + 335 pp.

Wesenberg-Lund, C. 1923. Contributions to the biology of the Rotifera. I. The males of the Rotifera. Kgl. Dansk. Vidensk. Selsk. Skr. Naturvid., 8, 4, N° 3: 191-345.

Wiszniewski, J. 1932. Les rotifères des rives sablonneuses du Lac Wigry. Wrotki piaszczystych brzegów jeziora Wigry. Note préliminaire. – Doniesienie tymczasowe. Arch. Hydrobiol. Rybactwa 6: 86-100, pl.III-IV.

Wiszniewski, J. 1934a. Les mâles des rotifères psammiques. Mém. Acad. Polon. Sci. Lett., Cl. Sci. nat., sér. B(II): 143-165, 1 pl.

Wiszniewski, J. 1934b. Wrotki psammonowe. Les rotifères psammiques. Ann. Mus. Zool. Polon. 10: 339-399, pl.58-63.

Wiszniewski, J. 1935. Note sur le Psammon. II. Rivière Czarna aux environs de Varsovie. Arch. Hydrobiol. Rybactwa 9: 221-238.

Wlastow, B.W. 1953a. European and North American rotifers of the Notommatidae family - symbionts on Cladocera and their species position. Zool. Zh. 32: 1110-1113. (in Russian).

Wlastow, B.W. 1953b. Interrelations between Cladocera and rotifers of the genus *Proales* living on them. Trudy Vses. hydrobiol. obsch. 5: 299-317. (in Russian).

Wlastow, B.W. 1954. Morphology and systematics of rotifers belonging to the order Monogononta. *Proales daphnicola*, a commensal of daphnids and related forms. Zool. Zh. 33: 50-64. (in Russian).

Wlastow, B.W. 1955. On the morphology of the males of some rotifers belonging to the order Monogononta. 1. *Proales daphnicola* Thompson and *Epiphanes senta* Ehrenberg. Zool. Zh. 34: 80-84. (in Russian).

Wlastow, B.W. 1956. Two new rotifers of the genus *Notommata* - *Proales lenta*, sp. n. and *Pleurotrocha larvarum*, sp. n. Zool. Zh. 35: 668-676 (in Russian).

Wlastow, B.W. 1959. Functional modification of bodyforms of Monogononta rotifer (Proalinae) and phylogenetic transformation of the shape of body in this group. Trudy Inst. Morph. Zhivtnykh Akad. Nauk. SSSR 27: 101-117. (in Russian).

Wulfert, K. 1935. Beiträge zur Kenntnis der Rädertierfauna Deutschlands. I. Teil. Arch. Hydrobiol. 28: 583-602.

Wulfert, K. 1937. Zur Kenntnis der Lebensgemeinschaften der Restlochgewässer des Braunkohlenbergbaues. 1. Die Rädertiere. Z. Naturwiss. 91: 56-69.

Wulfert, K. 1938. Die Rotatorien des Goldlochs bei Eisersdorf. Beitr. Biol. Glatzer Schneeberges, Breslau 4: 384-394.

Wulfert, K. 1939. Beiträge zur Kenntnis der Rädertierfauna Deutschlands. Teil IV. Die Rädertiere der Saale-Elster-Niederung bei Merseburg in ökologisch-faunistischer Beziehung. Arch. Hydrobiol. 35: 563-624.

Wulfert, K. 1940. Rotatorien einiger ostdeutscher Torfmoore. Arch. Hydrobiol. 36: 552-587.

Wulfert, K. 1956. Die Rädertiere des Teufelssees bei Friedrichshagen. Arch.Hydrobiol. 51: 457-495.
Wulfert, K. 1959. Rotatorien des Siebengebirges. Decheniana, Beih. 7: 59-69.
Wulfert, K. 1960. Die Rädertiere saurer Gewässer der Dübener Heide. II. Die Rotatorien des Krebsscherentümpels bei Winkelmühle. Arch. Hydrobiol. 56: 311-333.
Wulfert, K. 1961. Die Rädertiere saurer Gewässer der Dübener Heide. III. Die Rotatorien des Presseler und des Winkelmühler Teiches. Arch. Hydrobiol. 58: 72-102.
Wulfert, K. 1966. Rotatorien aus dem Stausee Ajwa und der Trinkwasser-Aufbereitung der Stadt Baroda (Indien). Limnologica (Berlin) 4: 53-93.
Wulfert, K. 1968. Rädertiere aus China I. Limnologica (Berlin) 6: 405-416.
Zavadowski, M.M. 1926. Rotifers of the family Notommatidae from the area of the hydrobiological station Zwenigorodskoj. Trudy Lab. Exp. Biol. Moskw. Zooparka, Moskwa 2: 261-295 (in Russian).

INDEX TO SPECIES

Page numbers printed in **bold** refer to main description ; page numbers in italics indicate illustrations. Entries of species names in parentheses are synonyms or species inquirendae.